Unfriending Dystopia

Unfriending Dystopia

RUSS WHITE

CASCADE *Books* • Eugene, Oregon

UNFRIENDING DYSTOPIA

Copyright © 2022 Russ White. All rights reserved. Except for brief quotations in critical publications or reviews, no part of this book may be reproduced in any manner without prior written permission from the publisher. Write: Permissions, Wipf and Stock Publishers, 199 W. 8th Ave., Suite 3, Eugene, OR 97401.

Cascade Books
An Imprint of Wipf and Stock Publishers
199 W. 8th Ave., Suite 3
Eugene, OR 97401

www.wipfandstock.com

PAPERBACK ISBN: 978-1-7252-7050-3
HARDCOVER ISBN: 978-1-7252-7051-0
EBOOK ISBN: 978-1-7252-7052-7

Cataloguing-in-Publication data:

Names: White, Russ, author.

Title: Unfriending dystopia / Russ White.

Description: Eugene, OR: Cascade Books, 2022. | Includes bibliographical references.

Identifiers: ISBN 978-1-7252-7050-3 (paperback) | ISBN 978-1-7252-7051-0 (hardcover) | ISBN 978-1-7252-7052-7 (ebook)

Subjects: LCSH: Computer Networks—psychological aspects. | Technology—Religious aspects—Christianity. | Theological anthropology—Christianity.

Classification: BT702 .W53 2022 (paperback) | BT702 (ebook)

VERSION NUMBER 09/30/22

Contents

Illustration | ix
Preface | xi
Acknowledgments | xiii

1 **Our Virtual Dystopia** | 1
  The Californian Ideology | 2
    Personal Autonomy | 2
    The Engineering Mindset | 3
    Progress | 5
    Naturalism | 6
    The Human in the Californian Ideology | 7
  Social Media and Beyond | 7

2 **Resonance and Goals** | 9
  Resonance | 9
  Goals | 11

3 **Flattening** | 14
  Self-Flattening | 14
  Interpersonal Flattening | 16
  Flattening at Scale | 17

4 **Impact** | 18
  Relationships | 18
  Freedom | 19
  Filter Bubbles | 20
  Responding to Social Media | 21

5 **Community** | 23
  Reject the Foundations | 23

        Politics and Freedom | 24
        Science, Engineering, and Freedom | 25
        Totalizing Life | 25
   Reject the Performance Mindset | 26
   Reject Surveillance | 26
   Reject Stand-Alone Virtual Community | 27
   Conclusion | 28

6 Don't Feed the System | 29
   The Importance of Data for Social Media | 29
   Learning Not to Share | 31
        Stop Recording Everything | 31
        Use Privacy-Focused Services | 32
        Use a Privacy-Focused Browser | 33
        Choose Small Things | 34
        Keep Things Local | 34
        Turn Things Off | 35
        Use a Virtual Private Network | 36
   Anonymization | 37
   If You Have Nothing to Hide . . . | 37
   Conclusion | 38

7 Refuse to Perform | 39
   Mental Resistance | 40
        Reject the Highlight Reel | 40
        Ignore the Likes and the Trolls | 42
   Practical Resistance | 43
        Minimize Notifications | 43
        Choose Intentional Interaction | 44
        Take Social Media Sabbaticals | 44
        Don't Filter | 45
   Concluding Thoughts on Resistance | 46

8 (Re)Learn to Think | 47
   Continuous Partial Attention | 47
   Shifting Perception of Truth | 49
        Immediacy | 49
        Factoids | 50
        Rejecting the Shift | 51

9 Christian Witness in a Social Media World | 53

Relationships | 53
The True Myth | 54
Facts | 57
Taking the Roof Off | 58

10 Unfriending Dystopia | 59
Unintended Dystopia | 59
Should We Just Disconnect? | 60
Unfriending Dystopia | 61

Appendix: Further Reading | 63
Bibliography | 67

Illustration

Figure 01: Data Flow in a Network | 36

Preface

MANY SCHOLARS—INCLUDING ECONOMISTS, POLITICIANS, psychologists, etc.—will say using social media is a net positive for you as a user and for communities at large. Others will say the negative impact of using social media outweighs the good.

Who is right? To answer this question, we need to go back to the culture questions.

Does technology shape culture, or does culture shape technology? Does the craftsman create the hammer, or does the person become a craftsman by shaping his muscles and thinking to the possibilities contained in the hammer? The answer is a bit of both; technologies come from within a culture and then shape the culture by creating and foreclosing possibilities.

One particular bit of technology captures and captivates our attention in the modern day—social media. The term *social media* encompasses not just services like Facebook or LinkedIn. While these are obvious forms of social media, there are many more, like your favorite online shopping site and the mapping (GPS) software you use to find your way around.

There are many questions we fail to ask about this technology. Where did social media come from? What culture did social media "grow up in"? Just as the size and heft of a common hammer changes the size of nails and the way we build houses, social media changes the way we build relationships, communities, and cultures. Social media impacts the way we construct ourselves—the way we see ourselves.

This book aims to answer these questions, beginning with the cultural source, moving through the cultural impact, and finally examining some practical ways you can control your use of social media so you capture much of the good while avoiding the bad.

Why am I the right person to write this book? I am not a philosopher (although I have some training in philosophy), nor a theologian (though I have some training in theology), nor am I an ethicist. In fact, I am a geek, a thirty-year veteran of the world of information technology. I grew up breathing the very air of engineering, and I have vast experience in everything from radios and radars through large-scale computer networks.

It is just this unique combination of skills that should convince you to read this book. I am a *fox* in knowing many different things relating to the problem of social media. On the other hand, I am a *hedgehog* in knowing this one thing—we are unique beings created in the image of God. This is the standard by which we should judge the culture of social media, both the culture it grew up in and the culture it creates.

I pray this book helps you understand social media and how to use it—and that you can take the lessons you learn here and apply them to many other areas of life.

Acknowledgments

I COULD, AND HAVE in the past, acknowledge those who have influenced my life, teaching me the things I know and helping me learn to think. Many people think intellectual achievement is something you "just have." They are wrong; thinking is a skill, like any other—a virtue to be developed across years of study and stubbing your toe against reality.

I would like to go a different direction with this book, however. The last few years of my life have been what many would call tragic; I have an ever-present, existential understanding of how culture can destroy a family and a person. Things I thought I had left in my childhood have reasserted themselves in ways I could not have imagined just a few years ago.

The joy of a life lived in a community of love, however, has worked (and is working) the kinds of healing only those who have experienced such a thing can understand. I would particularly like to acknowledge those who have, and do, believe in me through this journey, including my daughters, Doug Bookman, Tim Sigler, Mike Bushong, and my dear, sweet Adrianne.

I can do all things through Christ who strengthens me.

1

Our Virtual Dystopia

ON THE INTERNET, NO one can tell you are a dog.

Or what sex you are, what color your skin is, what your political beliefs are, or what your past is. This anonymity is the allure, the *promise*, of the digital world. You can be anything you want to be. And this is what was supposed to make the digital world a *new utopia*.

What about past attempts at creating a utopia? Didn't they all fail? This utopia would be different because it would be based on technology—previous failures were generally blamed on not having the right technology, rather than human nature—and . . . no one can tell you are a dog on the Internet. The somewhat anonymous—better described as *nonymous*—Internet environment allows people to recreate themselves as *better selves*. This freedom would allow people (and communities) to build a perfect version of themselves. People and communities would then be drawn into becoming the better version of themselves they have created online.

None of this worked. The digital world—social media in particular—inspires depression rather than joy, ideological bubbles rather than a free interchange of ideas, and flattening of the person rather than a broadening. Rather than a utopia, the digital world of social media and similar services can best be described as a *dystopia*. What happened? And more importantly—what can we do about this?

Answering these questions involves going behind the technology—understanding *why* social media works the way it does, rather than *how* it works.

The first three chapters of this book are designed to take you on a journey through *why* technologies like social media work the way they do.

The remaining chapters will make some practical suggestions about *how* individual Christians, the church at large, and anyone else concerned about the impact of these technologies on individuals and culture may respond.

The best place to start is the culture in which social media developed, the Californian Ideology.

The Californian Ideology

Four different streams of thought became prominent in the decades leading up to World War II (in the late 1930s through the early 1940s)—a strong sense of personal autonomy, the problem-solution engineering mindset, the idea of progress, and a firm rejection of religious belief in favor of naturalism. These four blended in the interaction between the hippie and military research communities in the 1950s, producing the Californian Ideology.

Personal Autonomy

A strong sense of personal autonomy can be traced to the religious wars beginning with the slaughter of one hundred French Protestants in 1562 at the hands of a Catholic duke. Both sides believed they held to the only true form of Christianity, and both were willing to die—and kill—in the name of their beliefs. The first efforts at peace, such as the Edict of Nantes in 1598, were notable because *governments were asked to resolve theological disputes.*

The idea that governments should be called on to resolve theological disputes, combined with technical advances leading to creating a "middle class" of skilled labor and business owners, led to the idea that religious belief is a personal decision. In 1688, the nobles of England, Scotland, and Ireland revolted against King James in the Glorious Revolution, codifying the right of individuals to choose their religious beliefs. This right of the individual to choose their individual belief developed into a right to privacy and then into the full-blown individual autonomy accepted as a foundational truth in modern culture.

Californian Ideology stretches personal autonomy to the next level. To be fulfilled, everyone deserves to become whatever they choose. Social media enables this goal by allowing users to present themselves as they want to be seen or known ("on the Internet, no one knows you are a dog"). As the user shapes their real life based on their chosen online identity (the intermixing of the real and virtual is why these are called *nonymous* rather than *anonymous*), the user progressively "becomes" the person they choose to be.

Social media also enables personal autonomy by overcoming the time and distance restrictions on relationships. Each user's community is the one they have chosen, rather than the one where they live.

The Engineering Mindset

Bringing people from the past into the present—just to see their reaction to how the world has changed—is a standard movie plot. What would a scientist from two hundred years ago find surprising about the modern world? Perhaps it would be the rapid advance in science or what humans have achieved in overcoming the physical limitations of the world around us. What they would probably be surprised at, however, is how little science there really is in the modern world.

The engineering mindset has largely overtaken science, making it into a practical affair. Much of this can be traced back to Francis Bacon's *new science*. Bacon argued that truth claims should be tested using experiments—the truth should be something you can see, touch, hear, taste, or smell. While human senses are imperfect, he argued instruments could be devised to aid the human senses, and experiments could be designed to be repeatable. Thus, Bacon said, experiments could be rendered the only reliable way to discover the truth.

Tying truth to physical tests ties truth to *practical results*. The surest way to prove a theory is true is to show it works "in the real world." This connection between truth and practical performance led science down the path towards an engineering mindset. Bacon's new science turns out, in many ways, to be a form of advanced engineering. The engineering mindset is: identify a problem to solve, brainstorm a set of solutions, test each solution, and then deploy the solution in the real world. Engineering is concerned with operational knowledge—how the world can be improved.

Social Media applies the engineering mindset to its users through *the nudge* and habit formation. *Nudging*, a term coined by Richard Thaler and Cass Sunstein, guides users through choices towards a goal.

> A nudge, as we will use the term, is any aspect of the choice architecture that alters people's behavior in a predictable way without forbidding any options or significantly changing their economic incentives. To count as a mere nudge, the intervention must be easy and cheap to avoid. Nudges are not mandates. Putting the fruit at eye level counts as a nudge. Banning junk food does not.[1]

---

1. Thaler and Sunstein, *Nudge*, 10.

For instance, automatically signing people up for savings through a retirement account unless they opt out and signing up people applying for a driver's license to donate their organs unless they opt out are examples of nudging a decision in a direction considered helpful or healthy. Social media nudges users in a thousand different ways, including—

- Suggesting users invite their friends to join the social media network.
- Defaulting to sharing everything; users must opt out of sharing rather than opting in.
- Making it easy to sign up for the service and difficult to close your account.

There are more controversial nudges built into social media, as well. For instance, Facebook once published a research paper showing how they could change the outcome of an election by encouraging some people to vote and *not* encouraging others to vote.[2] Robert Epstein and Ronald Robertson document the nudges given through search engines (called the search engine manipulation effect) can change the votes of hundreds of thousands of people.[3]

Some forms of the nudge are designed to encourage user behavior that is good for the company and *bad* for users—for instance, automatically adding extended download or ongoing subscriptions to a user's cart during the checkout process or making it impossible for a user to continue to content without clicking through to an advertisement. These are called dark patterns; they are quite common on social media services.

Social media also encourages users to form a habitual attachment to the service. User experience (UX) designers build these services around a behavioral plan, guiding users along a cue, routine, and reward path. The *cue* normally begins with external events. People get hungry during the day, remember a friend's birthday, or their phone (or watch) vibrates or plays a little tune. The goal of the app developer is to connect these external cues to internal ones—the user should want to check the app for some reason without external prompts. The movement from external to internal cues, repeated often enough, turns app use into a habit.

2. Sifry, "Facebook Wants You to Vote on Tuesday."
3. Epstein and Robertson, "Search Engine Manipulation Effect."

Progress

If you lived in the ages before the spread of Christianity throughout the Roman Empire, you would have probably seen the world through a lens of recurring time. Days, seasons, governments, and even the lives of gods and people all followed a circular pattern, repeating the same events across a defined timescale. However, the wide spread of Christianity imposed a new understanding of time; the world has a beginning and an end. There are patterns between these two, but not every season is precisely the same, nor every life.

This linear idea of time brings with it another idea—*progress*. If not every season of life is the same, these things can be improved over time. This idea of progress was difficult to square with everyday life up to the Industrial Revolution. Even though new ways of doing things were being invented, they didn't impact the daily pattern of life.

By the end of the Industrial Revolution, however, the idea of progress was hard to deny. Everyone could see the massive progress made by applying the engineering mindset. Airplanes, road systems, automobiles, widespread electrical power, sanitary water, waste systems, and modern medicine all attest to the massive progress made in turning nature into the servant of humanity.

World Wars I and II, by exposing the world to the apocalyptic results of technology developed and deployed to kill massive numbers of people, came close to destroying widespread faith in progress. It was difficult to look at images of trench warfare, the remains of cities destroyed by firebombs and nuclear blasts, and the chilling productization of death in the Holocaust and see progress.

The computer revolution, flowering in the 1960s, however, gave believers in progress new hope. Here was a technology that could be used to improve the efficiency of almost anything to which it was applied. It could also be used to create a virtual world—a world built on the highest ideals of the most intelligent and caring. Here is a technology that can shape individual users, improve their lives and the world itself through these users.

The high-minded goals of social media can be seen in the mission statements of the companies that create and operate these services. LinkedIn's vision is to create "economic opportunity for every member of the global workforce" by connecting "the world's professionals to make them more productive and successful."[4] LinkedIn seems to fulfill this role by being a

---

4. "About LinkedIn."

source of leads and customers for businesses and a source of professional talent for recruiters.[5]

Google proclaims it is "committed to significantly improving the lives of as many people as possible."[6] Earlier Google mission statements included "to organize the world's information and make it universally accessible and useful," and the ever present "don't be evil." Go to Silicon Valley, mingle with anyone working at any of the large social media companies, and ask individual employees what their goal is. They will universally tell you they are working to make the world a better place.

Naturalism

While naturalism and atheism are closely related, wide-ranging belief systems, all their forms have one core: a strong anti-theism. C. S. Lewis says this anti-theism holds that the "ultimate Fact, the thing you can't go behind, is a vast process in space and time which is going on of its own accord."[7] According to Carl Sagan, "[t]he Cosmos is all that is or was or ever will be. Our feeblest contemplations of the Cosmos stir us—there is a tingling in the spine, a catch in the voice, a faint sensation, as if a distant memory, of falling from a height. We know we are approaching the greatest of mysteries."[8]

The culture of Silicon Valley, where social media systems were first conceived and built, is intensely naturalistic. While many are not strictly atheistic, they firmly believe there is practically nothing outside nature. If there is any sort of god or gods, these beings have little interaction with the physical world. Humans have been left, forlorn, to either build a bright and glorious future on our own—or to fail on our own.

One of the practical results of this naturalism is that *humanity is a part of nature*. Humans do not have some sort of dignity imparted through divine creation, nor a fallen nature that must be corrected through supernatural means. Instead, humans are simply to be studied and understood much like any other part of nature—and ultimately shaped, so each individual is more useful for society at large. As C. S. Lewis says, "what we call Man's power over Nature turns out to be a power exercised by some men over other men with Nature as its instrument."[9]

---

5. Osman, "Mind-Blowing LinkedIn Statistics and Facts (2019)."
6. Google, "Google's Mission, Values & Commitments."
7. Lewis, *Miracles*, 7.
8. Sagan, *Cosmos*, 1.
9. Lewis, *Abolition of Man*, 55.

The Human in the Californian Ideology

These, then, are the components of the Californian Ideology—personal autonomy, the engineering mindset, progress, and naturalism. How does this ideology see the person?

A person should desire to—and should be—shaped to fit into an increasingly improved future. The person's value does not come from some external source but rather from how they fit into and improve the community in which they live.

There are several inherent contradictions in this view of the human. How can each person be fully autonomous (the community should respect their decisions on who they are and want to be) and yet be shapeable for progressive means? How can a person's value be tied to their community, yet they are expected to choose their community freely? These contradictions are, in fact, fatal to the Californian Ideology, but they are often covered by a strong dose of "magic smoke" emanating from the technological world. These will prove to be good points of attack against this worldview, but for now, we need to simply accept the Californian Ideology as it is to fully understand the results of social media in the real world.

Social Media and Beyond

Before moving forward, it is helpful to reflect on the scope and nature of the kinds of technology discussed here. *Social media* is the common name for this technology—and the one used throughout this book (for simplicity). There is a more technical name for this kind of technology, however—*neurodigital media*.

This kind of technology is *neuro* because it is designed to interact with the user's mind. The world created by neurodigital media is *virtual*; it exists as wires, transistors, and electronic bits, but our mind interprets these things as a world. This world is almost entirely centered in the mind.

This kind of technology is *digital* because it relies on high-speed digital computing technologies. While these technologies were originally invented in the 1950s, the scope and scale of these digital systems have expanded until they are almost ubiquitous.

This kind of technology is *media* because we interact with it in ways similar to older media. We watch, read, talk, and write on social media.

While social media might be the most obvious example of neurodigital technology, many other systems fall under this description. Large-scale shopping sites, such as Amazon and Flipkart, rely on many of the same surveillance, analysis, and user interface techniques as all the social media sites do. Dating apps do, as well. Even services that might not seem obvious candidates like mapping services and search engines are, in some sense, neurodigital media.

While the scope of these technologies is broader than *social media*, they will all be called social media throughout this book (to avoid confusion). Social media, then, is not just services like Facebook and LinkedIn—it is a collection of services that use similar techniques and have a similar view of the human person.

2

Resonance and Goals

THE PHILOSOPHICAL UNDERPINNINGS OF social media—the emphasis on autonomy, the engineering mindset, progressivism, and naturalism—are an important beginning. This chapter will move beyond these underpinnings and discuss other topics that will help determine how individual believers and communities should react to social media.

The discussion will begin with the resonance of social media, or what can be considered the natural way social media services present information and how users engage with one another over social media. While the first chapter considered some of the more idealistic social media goals—building a virtual utopia—this chapter will discuss the financial goals of the companies that run social media services. How do they make money, and what does that imply about how they treat users?

Researchers and writers spend a lot of time thinking about the impact using social media has on individuals and communities; the next section is a short overview of this research. Finally, we consider a general strategy for managing (or combatting) the impact of social media for Christians and Christian communities.

Resonance

When you began reading this book, what did you expect to experience? You were probably wondering about the impact of social media and how to manage its effects in your life or community. Some books, then, can be focused on *providing information*—although most agree that information is best provided as part of a story or formatted as a logical argument. Both things—telling stories and various forms of logical arguments—are forms of *narratives*. In both cases, there is a sense of "flow" or "movement."

Another way to put this is *to convey information effectively in a book, you should turn it into some kind of narrative*. Writers can communicate with readers by converting whatever material they have into some kind of narrative, whether it's a narrative through time or a narrative through ideas, and whether it's a true account of things or not (fiction—although fiction is more effective when it describes a truth about the state of humanity).

A common medium is television. There are two sides to television—what happens behind the screen and what happens in front of the screen. Behind the screen are actors, musicians, talking heads, producers, and others who make this virtual world seem real (the music group Farrell and Farrell call these "People In A Box"). In front of the screen is an audience.

Almost all communication in the world of television is one way, from the performers to the audience. There may be the occasional chance for the audience to vote, buy, or otherwise provide feedback, but the audience is anonymous to the performers. The performers entertain, the audience is entertained.

Neil Postman says television "has made entertainment itself the natural format for the representation of all experience."[1] Performers can best communicate with a television audience by converting every kind of available material into some form of entertainment—including such things as the news, the weather, worship, and preaching.

This natural way of communicating—the form of communication a media turns every kind of message into—is called the *resonance*. Neil Postman says:

> Every medium of communication, I am claiming, has resonance, for resonance is metaphor writ large. Whatever the original and limited context of its use may have been, a medium has the power to fly far beyond that context into new and unexpected ones. Because of the way it directs us to organize our minds and integrate our experience of the world, it imposes itself on our consciousness and social institutions in myriad forms.[2]

The resonance of every media is drawn into the real world by setting our expectations.

Those living in a reading culture will expect expert storytelling and arguments with a logical flow. To be brought before the throne of God in a

---

1. Postman, *Amusing Ourselves to Death*, 87.
2. Postman, *Amusing Ourselves to Death*, 18.

reading culture is to encounter God through thought and the stories of these great acts in human history. Those living in a television culture will expect to be convinced through entertainment. To be brought before the throne of God in a television culture will require drama and special effects—to be entertained by the presence of God.

If the resonance of writing is the story, and the resonance of television is entertainment, what is the resonance of social media? Social media is much like television because there are two sides of the screen: the performer and the audience. However, the primary difference is that *everyone is on both sides of the screen* in social media.

Which is more attractive to you—to be a part of the audience or the performer? Think of every child of a certain age—if they could do anything, what is it they would do? For instance, when a child says, "I want to be a fireman," what is the child thinking? They are thinking of the act of being a hero—performing the acts of a hero for all to see—so they can be seen as good, strong, and worthy persons. Just about any dream a child gives probably revolves around this idea of being seen as a hero or of having an audience.

It is innate in every human to want to perform—to be seen, known, liked, and accepted. Social media offers the entire world as a stage to every user—the temptation is just too much for most people to ignore. The offer is close to one of the three temptations of Christ in Matthew 4:8–10 (ESV): "Again, the devil took him to a very high mountain and showed him all the kingdoms of the world and their glory. And he said to him, "All these I will give you, if you will fall down and worship me." The average social media user doesn't have the spiritual skills or life wisdom required to refuse this kind of offer.

Because the desire to perform is stronger than the desire to be entertained, performance takes the upper hand—social media's resonance is *performance*. In the virtual world created by these technologies, everyone is performing to everyone in the world.

Goals

Social media systems are complex to build and maintain. To give a rough sense of the scale of the systems required, consider—

- Facebook is reportedly building or expanding three different data centers in 2021, encompassing 9.7 *million* square feet of space and costing at least $50 billion.[3]
- It is estimated that Microsoft operates over 1.6 million computers (servers) to support its various services.
- Even a "smaller" social media service like LinkedIn operates over 200,000 computers (servers) in globally located data center facilities.

These large data centers and networks are expensive—where does all this money come from? While some services, like LinkedIn, have premium tiers of service, most of this money is earned by selling advertising and information about users.

Social media services must capture the attention of millions of people to reach them with advertising. Social media services must keep users engaged to capture information about users valuable enough to sell. Each time a user "likes" some bit of information, "reshares" with other users, connects to another user, posts a picture of a meal they just received at a restaurant or that beautiful sunset they just saw—each and every action each user takes—data is generated.

This mountain of data can be analyzed to produce what might be called a profile on each user. Those who have delved into the information their profiles contain often find the experience harrowing. Whether or not the information is always correct, it is always detailed to a level few expect. There doesn't seem to be much harm in a company like Facebook knowing a lot about you—so what if they know what your favorite color is or what political party you normally vote for?

The main goal, however, is prediction.

The service provider can predict what you will do when presented with specific choices or situations by collecting and analyzing this information. While these predictions are not always accurate, combining knowledge of what party you will vote for and what kinds of prompts will cause you to vote can be powerful.

Perhaps the party you normally vote for will pay the service to remind you to vote in just the right way, or perhaps the party you don't normally vote for will try to convince you to change your vote—or perhaps the party you don't normally vote for will pay the service *not* to remind you to vote. Shoshana Zuboff says a "marketplace for behavioral prediction" developed,

---

3. Lundak, "$1.5 billion Facebook data center"; Gattis, "Facebook Spending Tops $1 Billion"; Schilling, "Facebook Planning $42 Billion Data Center Expansion."

which trades in what she calls "behavioral futures."[4] Selling influence developed through surveillance is what Zuboff calls surveillance capitalism.

Social media services do not sell all the information they have about you, however. Their power rests in capturing your attention, which means they want you to check their service—to be engaged in their service—habitually. While the habit-forming techniques described previously are effective in a general sense, tuning them for each user makes them more powerful.

It turns out that the more you know about your users, the more effectively you can consume their attention. Knowing more about your users leads to what Moshe Vardi calls Kai-Fu Lee's virtuous cycle: "More data begets more users and profit, which begets more usage and data."[5] Thus we return to the desire to increase user engagement—to draw users in, make them pay attention more deeply and more frequently. The more a social media service can seep into every corner of every user's life, taking those spare moments while you're waiting in line (rather than chatting with your spouse or the cashier), the more successful the social media service will be. Social media can consume so much of our lives that it reaches the point of appearing to be obsessive—or even insane. Reed Hastings claimed in 2017 that sleep is the primary competition for the Netflix streaming video service.[6]

---

4. Zuboff, *Age of Surveillance Capitalism*, 7.
5. Vardi, "Winner-Takes-All Tech Corporation," 7.
6. Raphael, "Sleep Is Our Competition."

3

Flattening

You probably fold or *flatten* boxes to store or recycle them—you reduce them from a three-dimensional object to something closer to a two-dimensional object. Flattening can also be applied to people by removing their wholeness or completeness. Readers of fiction often complain about "one-dimensional" characters in a book—they show no depth or nuance.

Can you flatten *people?* Yes. When you visit a doctor, they primarily consider you a body rather than a whole person. Whether you like to water ski is not important to a doctor interacting with the patient unless it impacts their physical health in some way. Police officers stopping a motorist also flatten a person. It's not important *why* you were speeding, only that you were.

There are circumstances when it's ethically okay to flatten someone or see only some of the aspects of the whole person. Social media, however, encompasses all of life, from career to relationships to belief. When social media flattens a person, the flattening effect can seep into real life in harmful ways. There are three ways social media flattens people: users self-flatten, users flatten one another, and the social media network itself flattens users.

Self-Flattening

You probably see it hundreds—perhaps thousands—of times in a week. A small group of people, usually teenagers, or a single teenager, will stop, "put on a face," and take selfies. What is the selfie-taker saying about themselves? Perhaps they are saying "I was here," laying down a marker of presence in time like those little piles of stone often built in faraway places. Or perhaps they are saying, "this is how I want you to think I live my life, the way I look, who my (cool) friends are, *who I am.*"

Is the person taking the selfie giving a true picture of themselves? Pictures can never show the whole person, and even video is restricted to representing a real person in the moment.

Does the selfie-taker seriously want to show themselves as they truly are? To ask the question is to answer it—clearly, the answer is no. The selfie must be perfect. For those without a perfect face or body, thousands of filters can reshape the image to be closer to what they want to present. AirBrush can remove wrinkles, whiten teeth, and enhance "problem areas." Facetune2 auto-detects different areas of the user's face and allows you to "change the curve of your smile, the size of your jaw, the width of your face, resize features, use filters, and more."[1] Women tend to slim down their faces, make their faces rounder, and increase the size of their eyes using these kinds of filters. YouCam Makeup allows users to add a perfect makeup job to their selfies, and Retrica creates an edgy, "old-fashioned" look.[2]

The selfie is a kind of performance. The selfie-taker is putting on an act for an unseen audience, hoping they are seen and liked. The user is presenting a hoped-for possible self rather than who they are. A self who will be looked at, watched, followed, and liked. Jeremy Rifkin says, "Growing numbers of people, especially young people, see themselves as performance artists and their lives as unfinished works of art."[3]

The selfie is only one kind of performance social media encourages, however. The feed created by posts to Facebook is another kind of performance—users seek to show the world they live the perfect life, dwelling in the perfect house with the perfect family and community. It's rare for someone to show their life as less than perfect on a platform like Facebook—and if they do, it's often to illustrate their status as a victim of some cosmic injustice.

LinkedIn is no exception, either. Users spend hours polishing their profiles to attract attention because attention leads to better positions, better pay, and a better life. Companies polish their LinkedIn presence, as well, hoping to attract the highest level of talent possible.[4] Because the resonance

1. Grasso, "10 Best Apps for Shooting and Editing Selfies."
2. Grasso, "10 Best Apps for Shooting and Editing Selfies.".
3. Rifkin, *Age of Access*, 213.
4. For instance, one senior manager at a social media company noted they try to make their technology presentations as "cool as possible," because such presentations are the backbone of strong recruiting efforts.

of social media is performance, everything on social media tends towards being a performance of some kind.

These performances of a hoped-for possible self are a form of flattening. The person presented online is a two-dimensional version of the whole person—it is shaped and formed to leave a specific impression, to gain a particular reaction. It is more like a character in a televised drama than we might be willing to admit.

If they stayed online, these performances would be fine, but social media is not a completely anonymous space. There is some overlap between a user's relationships on each platform and their relationships in real life. Because of this overlap, Zhao and colleagues call these environments nonymous.[5] In nonymous environments, the user can freely shape their online identity, which overlaps with their offline identity and then validates their identity in both online and offline settings.

Just as the idea of entertainment has changed the acts of preaching and worship, so too is the idea of performance changing the individual lives of everyday life. The resonance of the virtual world creeps into the real world by setting expectations that people must fulfill if they want the attention of any sizeable audience.

In performing, users self-flatten; this self-flattening all too easily transfers from the virtual world to the real one.

Interpersonal Flattening

When you post something on a social media site like Facebook or Twitter, what are you hoping will happen? Would most users be happy with just seeing it on their timeline? Most users post to inform, or entertain, or make a point—but in all these cases, the goal cannot be served unless other users pay attention.

How do you know if other users are paying attention or the size of your audience? Because you can tell how many people look at, like, dislike, comment on, or otherwise interact with your content. Measuring a post's performance is called quantification.

The quantification process moves easily from the post to the relationship. Other users are flattened merely into a quantifiable audience rather than seeing other social media users as individuals—people who might be having a bad day or good, a whole person with joys and cares of their own.

And once the audience can be quantified, our all-too-human desire for more takes over. The audience becomes a way to obtain more—rising

---

5. Zhao, Grasmuck, and Martin, "Identity Construction on Facebook," 1818.

numbers mean social approval—rather than individual *people* we can relate to personally. The old public speaking tricks of looking at no one in the audience, or looking at one person all the time, become the daily habit of life *with everyone.*

Some social media systems, particularly in the sharing economy, allow users to rate other users directly. When you catch a ride in an Uber, you are sitting in a car with a very real person—but this person will be rated by you and will rate you in return. What would normally be a pleasant human interaction is turned into a quantified struggle for the best rating possible by both driver and rider. Quantifying other people in this way is acidic, destroying the ability to create and build genuine person-to-person relationships.

It is much easier to flatten other users into *merely* an audience in social media because of the digital nature of these platforms. Email, for instance, is notorious for lacking the kinds of clues necessary for readers to capture nuances of expression often found in face-to-face communication. Even given the prevalence of emojis, stickers, and other methods of expressing emotion, social media focuses content on short bursts of information that often lack the context and nuance of other forms of communication.

Flattening at Scale

If the service is free, you are the product. Social media services treat their users as a product. These companies surveil, analyze, and test each user until they can be confident in their ability to predict each user's decisions in various situations. The ability to predict and guide users is what these companies sell.

The social media company must flatten users at every step in the process—from gathering information to selling a user's attention. Users are two-dimensional *information sources* rather than *individual humans* with God-given dignity.

4

Impact

THE FIRST CHAPTER EXPLAINED the Californian Ideology, which holds that humans are a shapeable (or manipulable) part of nature, and humans should be shaped to produce progress through history—that social media makes it possible for individuals and cultures to be on the right side of history. The second chapter considered the resonance and goals of social media, and the third discussed the idea of flattening people through social media. It's time to bring all these elements together and look at the impact of social media on individual thinking, relationships, and freedom.

And then to begin looking at how Christians, Jews, and others who hold to a more traditional view of people, culture, and society, can respond to the modern-day flood of social media.

Relationships[1]

Social media is supposed to bring us closer to friends and family by creating a virtual space in which relationships can grow, and yet we perceive those who live primarily in virtual spaces as outcasts of a sort. Which of these two is true—is social media good for relationships or bad? Both. Social media is good because it does create a feeling of constant contact with family and friends.

Social media is bad because—without a strong, regular dose of "in real life" contact—it flattens relationships into facts, which are then quantified

---

1. For those interested in empirical evidence, an appendix of studies showing the negative effects of heavy social media use can be found at the end of this book.

through likes and reshares. God made humans to live and thrive in and through relationships. Quantified relationships cannot fulfill the human need for deep, meaningful human contact. Is it any wonder heavy social media use encourages loneliness, including the loneliness that leads to depression?

The quantification of relationships also allows us to compare ourselves more easily to others. Each time we see a connection's post about a new certification, a new position, their children receiving some prize, or anything else strongly positive, we find ourselves feeling a little more like we are missing out on important things.

However, most people are performing on social media—what they post and say is a highlight reel of their real life.[2] Comparing your all-too-real everyday life to the apparently perfect life others are performing on social media is a depressing habit.

Freedom

The nudge, the habit cycle, and addiction fall on a continuum of reducing human freedom.

According to Richard Thaler and Cass Sunstein, the nudge is a "[self-conscious attempt] to move people in directions that will make their lives better"[3] by architecting choices. When a social media service emphasizes that other users voted in your timeline, they nudge you. Another way to architect choices is by making "better" choices more prominent in a user interface, such as when a bank encourages a customer to transfer money to a savings account while they are in the process of depositing a check.

The mirror image of the nudge is the dark pattern. Instead of encouraging users to move people in directions that will improve their lives, dark patterns attempt to move people in directions that increase corporate profit (or control). Dark patterns include placing an "extended download fee" in a user's shopping cart or automatically adding a subscription to an advertiser's mailing list. There is no obvious way to "opt out" of these additional services the user did not ask for in some cases. The only difference between the nudge and the dark pattern is the intent of the designer.

Both nudges and intent can reduce user freedom by filtering choices. Whether the intent is to improve users' lives or increase profits (most often,

---

2. Steers, Wickham, and Acitelli, "Seeing Everyone Else's Highlight Reels," 728.
3. Thaler and Sunstein, *Nudge*, 10.

in Silicon Valley, these are seen as complementary goals), choices are controlled in a kind of libertarian paternalism.[4]

User experience designers also try to form a usage habit in users. Social media sites make money through engagement or getting users to pay attention to the service as often and long as possible. Stephen Wendel outlines four steps to designing a product that can change user behavior: *understand* how humans decide, *discover* the right behaviors to change, *design* the product around the behavior to be changed, and *refine* the impact of the product on behavior.[5]

While these four steps can exist in physical product design, they are greatly enhanced in social media. User actions can be recorded (surveilled) in detail, so designers can understand how each user decides. Varying choices can be presented to each user over time, allowing designers to refine the user interface, increasing the impact of the design on decision-making. As Zuboff says, "Surveillance capitalists discovered that the most predictive behavioral data come from intervening in the state of play in order to nudge, coax, tune, and herd behavior toward profitable outcomes."[6]

Addiction is the most extreme case of reduced freedom. An addiction is any process over which we feel powerless—the decision to take some action is completely removed from our hands and placed under the control of internal or external triggers. There is some controversy over whether social media is addictive. While social media is not addictive in the same sense as alcohol or other substances, it does have addictive properties. Some people can become addicted to social media use in much the same way they become addicted to gambling. For instance, studies show social media can impact users' mood,[7] and users sometimes experience withdrawal symptoms when quitting social media use.[8]

Filter Bubbles

You might like or share something because you find it funny, inspirational, or important. Other users, however, are not the only ones watching what you read and share on social media networks—the network itself is watching. What can the network operator learn about you by watching what you

4. Thaler and Sunstein, *Nudge*, 8.
5. Wendel, *Designing for Behavior Change*, loc. 309.
6. Zuboff, *Age of Surveillance Capitalism*, 8.
7. Tams, Legoux, and Léger, "Smartphone Withdrawal," 1.
8. Rosen, Cheever, and Carrier, eds., *Wiley Handbook of Psychology, Technology, and Society*, 472.

read and share? The operator can learn what emotionally engages you and what will keep you on the service longer and bring you back for more. Once the service uncovers what will keep you engaged, it will keep feeding you more of "that kind" of content—whether it is cute kittens or over-the-top political discourse.

What you engage with, then, controls what the social media service places into your feed. The more you use the service, the less you will see things you disagree with or things that at least make you think. The social media service itself becomes a personalized echo chamber for ever-smaller groups of users and individual users. What you engage with builds a filter, and the social media service puts you into a filter bubble.

The distorting effect of filter bubbles on society can hardly be overstated. People gather into ever-smaller groups, each with a completely different view of reality—and each completely convinced their view of reality is "all there is to know." Groups develop and rely on different news sources that will bend any situation to fit into the group's narrative because this is the way to gain and hold attention. Society is torn into small groups, using completely different sets of "facts" and talking past one another rather than to one another.

Once these filter bubbles form, the social media operator itself can control the discourse by allowing some groups to spread and refine a narrative that fits into their view of the world while filtering out (or simply delaying—information's value varies over time) the version any other group might put forward. Through the simple act of filtering, social media services can control much of society's political and cultural discourse.

Responding to Social Media

Understanding how and why social media produces its effects on individuals and cultures allows us to build a road map to fight back. How can we counter this flood?

Begin at the beginning with the Californian Ideology. Reject the foundational idea that humans can, and should, be shaped. Reject the shallow vision of progress that declares one set of ideas on the "right side of history." Reject naturalism and its view of humanity as just another part of nature. These counters are all best accomplished through communities and hence we will discuss how communities can counter the effects of social media.

We must each find ways to disrupt the performance and fight against the nudge at an individual level. Each person will need to build on and modify the general suggestions given in the remainder of this book. Everyone's ability to resist addiction, or perceived level of privacy, will necessarily be different. There are no right (or wrong) answers, but the answers must be intentionally designed to counter some aspect of how social media impacts your life.

Three general strategies will be discussed in the following chapters. First, *don't feed the surveillance machine*. Second, *refuse to perform*. Third, *counter social media thinking habits*.

It's important not to pretend these things are easy. Countering the harmful effects of social media in our lives will mean undergoing some inconvenience—and potentially even missing out on some financial or life opportunities. But if we fail to work against these effects, we could lose the most significant opportunity of all—the chance to fully become the humans God intended us to be.

5

Community

While individuals use social media, communities—like churches—can help individuals manage their presence and use of social media. Community leaders can teach about how social media impacts lives using the first four chapters (and the additional resources listed in the appendix) as a guide.

The sections below outline other ways a community (and community leaders such as pastors) can reduce the adverse effects of these technologies, beginning with rejecting the foundational worldview embedded in social media. Philosophy and philosophical commitments might seem like a strange (or even heavy) place to start, but attacking the reductionistic view of the person is critical. The ills resulting from the animal-focused view of the human are not restricted to social media.

Reject the Foundations

The modern era turns all the dials, so they point in one direction—towards personal autonomy. The very concept of progress is tied up with making the individual freer—specifically in the sense that each individual can make and remake themselves as many times as they like across their lives, doing and being whatever they like. This radical individualism serves as a critical foundation stone of the virtual world envisioned by social media's creators. Communities need to reject this radical individualism in politics and science instead and focus on a more balanced view of the human person.

Politics and Freedom

Politics was originally aimed at building communities in which individuals could live full, healthy, prosperous lives—where people could flourish. Building such communities, however, meant setting community standards. It meant examining what works in the real world, what makes the community stronger, and what does not.

Communities are a two-edged sword—they support, but they also restrict by expecting members to support other members, believe certain things, and act certain ways. The restrictions of living in a traditional, supportive community are often seen as reducing freedom—often translated into reducing dignity. How can individuals be fully "true to themselves" if they live within a community that restricts their beliefs and actions in meaningful ways?

A biblical account of freedom resolves this dilemma by noting there are two kinds of freedom: *freedom from* and *freedom to*. A prime example of the difference between these two is the story of Isaac, beginning in Genesis 16. God's promises to Abraham about Isaac seem to directly contradict one another—that his family would eventually be taken into slavery, where they would stay for four hundred years. Yet, Isaac's family would become a great nation. Why would God put a people he planned to make into a great nation through four hundred years of slavery? To be free to create a great nation, Israel had to learn to be a great community. Those four hundred years of slavery taught Israel how to be a community—and hence how to be free to create.

Instead of encouraging community, the modern political mindset focuses on maximizing individual freedom, even to the point of covering over the consequences of a lifetime of poor decisions and forcing into silence those who believe there are external moral standards.

These values—maximizing individual freedom politically and economically—are part of the ethos or vision of reality of social media. Every change in law or economics allowing the individual to become more autonomous, from welfare to easy divorce, is cast as progress.

We (as communities) should reject this radical focus on individual autonomy in every arena of life. Rejecting a focus on individual autonomy does not mean rejecting personal autonomy entirely but rather learning how to balance individual autonomy and community.

Science, Engineering, and Freedom

Science and engineering began as quests to shape nature towards human flourishing. Both, however, have shifted to focus on creating radical human autonomy by overcoming nature. Progress in science and engineering is anything that allows humans to overcome nature in some way.

The science of human psychology has been turned to the nudge and modern propaganda techniques to manage entire populations. For instance, some governments have discovered they can engineer the beliefs of an entire population by injecting false information alongside the truth.[1]

The science of human physiology began with learning how to heal, and then increased its scope to improving human flourishing through nutrition and managing environmental factors. More recently, however, they have turned to supporting radical individualism in childbearing and altering our bodies to fit what we feel about ourselves.

These values—maximizing individual freedom by overcoming nature—are part of social media's ethos or vision of reality. Every discovery that disconnects the person from the body is celebrated widely.

We (as communities) should reject the use of science and engineering to seek radical autonomy. Rejecting the use of science and engineering to seek radical autonomy will likely be called "anti-science"—but there is deep irony in this. We are told we must live closer to nature for environmental reasons but that we must also reject our human nature to live radically autonomous lives. The ideal of radical autonomy is not scientific but rather a philosophical commitment imported into science.

Totalizing Life

A more subtle effect of the nonymous nature of social media is its totalizing tendency. Totalizing is the "flip side" of flattening; the box, when folded, becomes flatter in some dimensions and larger in others. In the same way, the narrow performance—whether political, sexual, or social—becomes the goal of the performer's entire life. The totality of the person's life becomes the political, the sexual, or the social.

There is a temptation to see this problem as a political one—with people on the right totalizing their lives around their nation, a flag, or some other issue. The totalizing, however, is a danger across the entire ideological spectrum. While people on the right can immerse themselves in their national identity, people on the left can immerse themselves in their sexual or

---

1. Pomerantsev, "Beyond Propaganda."

political identity. Everyone, with enough focus, can focus their lives so narrowly on a single cause that they lose sight of the humanness of those they agree with, those they disagree with, and even themselves. Both right and left extremes must be rejected in favor of a strong community combined with well-defined and strongly enforced individual rights.

Faith leaders must strive to balance the poles of radical autonomy and immersing the person wholly in a community.

Reject the Performance Mindset

Today, the average person walking into an average church service is often greeted by a well-dressed person (sometimes chosen for their attractive physical appearance), guided to a large auditorium featuring a stage with well-designed lighting. The service often begins with music most would have considered professional grade in years past. The centerpiece of the service is a professionally preached sermon—sometimes purchased from a sermon preparation website (or with the aid of professional sermon creation tools), or even developed "after" some other (famous) pastor's sermon.

But before you single out large "seeker-friendly" churches for their style of service, even Orthodox and more traditional churches—and even synagogues—are often focused on performance. Instead of loud music and fancy lights, there are special vestments, robed choirs, and an order of service designed to impress.

Some churches have moved towards "playing to the camera"—building their service, lighting, and music around creating a visual presentation suitable for video (and social media).

There is nothing wrong with a fine performance—of course. What is a problem is focusing so deeply on the performance that you lose sight of the *purpose* of a church service—training individual Christians. Worship and teaching can be entertaining, but they are not entertainment. Churches need to reject the resonance mindset, thinking deeply about the meaning and purpose of music and sermons in the life of Christians.[2]

Reject Surveillance

Social media companies make their money by selling information and attention—information about their users and the attention of their users.

---

2. Works such as Schaeffer, *Art and the Bible,* and Begbie, ed., *Beholding the Glory,* are great resources for pastors and worship leaders who want to think deeply about the use of music and visual arts from within a Christian worldview.

They grow through Kai-Fu Lee's *virtuous cycle;* more information can be used to create a more customized experience. A more customized experience causes users to engage more fully, and more fully engaged users generate more information. The next chapter will consider how users can act to break this cycle at an individual level, but communities have a role to play as well.

Asking people to shut off their cell phones during service, for instance, can help people focus their attention on spending time in community, worship, and teaching. Churches and other organizations can also organize events where cell phones and other electronic devices are not permitted. Sports ministries, small groups, and even weekly prayer and study meetings could greatly benefit from activities focused on face-to-face "IRL" time.

Security cameras and other security systems are a more controversial area where churches can act to reduce information being fed into the surveillance capital system. Having video cameras can help see what's going on in a physical area and even help catch those who might damage the facility in some way—or to catch interactions that simply should not happen. Churches, however, might want to resist the draw towards "cloud-based" monitoring systems. They are simpler to set up than an in-house system. Still, automated video analysis tools (such as facial recognition), cameras, and other security systems can become an unintended source of information on community members.

There are other places community leaders and members can look to help reduce surveillance, but it takes some creativity to think through how things might be used rather than you intended.

Reject Stand-Alone Virtual Community

Remote work has been growing in scale and scope for many years—I began working remote part-time more than twenty years ago, and over the intervening years I have transitioned to full-time remote work (plus a moderate travel schedule). Over those twenty years, many knowledge workers have moved to at least part-time remote work. The global pandemic beginning in early 2020, enabled by the gig economy, rapidly accelerated remote work to the point where many people decided not to ever go back to a physical office.

The impact of shutting down thousands of schools had a similar impact on schools. Tens of thousands of parents, having experienced homeschooling for the first time, have realized the flexibility and family "together time" far outweigh the value of in-person schooling.

While it's hard to tell if these trends will ultimately reshape our cultures and societies, we can be certain the level of comfort with online interactions continues to increase. People will be doing more "virtually" and less in person (or IRL).

It's tempting for churches to follow along, transitioning to an "online-first" stance for studies, services, counseling, etc. How long before the "first church of Facebook" is established (there might already be such a thing), an entirely virtual church?

Churches should resist virtual-first community, no matter how tempting it might be. Physical, in-person interactions carry much more depth. It's not just the cues interchanged through facial and bodily expressions—touch, talking in person, and worshipping as a body in person are all important parts of the human experience.

It is good for churches and other communities to use virtual spaces of all kinds to *augment* in-person, physical gatherings. But we should reject *replacing* physical gathering with virtual gathering.

Conclusion

We have a modern tendency to believe progress is always pointed towards individual autonomy (not freedom, but autonomy—these are different things). But is this progress? Is it progress to replace physical church services with online services and relationships built sitting over a cup of coffee with relationships built sitting over a video camera?

The state and science cannot create individual freedom of the kind we seek; God has created the world in a way that only allows freedom and flourishing in community and within the guardrails he created. Cultures where there is nothing other than the state and the individual are necessarily totalitarian.

6

Don't Feed the System

COMMUNITIES HAVE A DEFINITE role to play in reducing the influence of social media in the daily lives of their members—but much of what communities can do is support individuals. Communities can teach individual members about the dangers of social media, and build cultures that encourage members to use social media rationally. What kinds of things can individuals do to reduce their exposure to the power of social media?

The first of these is *don't feed the system*. This chapter will begin by reviewing the importance of intimate surveillance to making these systems work. While you cannot truly "go dark," there are ways you can reduce the breadth and depth of surveillance in your life.

The Importance of Data for Social Media

Every social media system operator will gladly tell you what their purpose is. Facebook is designed to keep you connected to friends and family. TikTok is designed to entertain you with short videos. Amazon is designed to simplify your life by unifying the shopping experience for everything in life. Google search is designed to help you find what you want or need.

All of these stated purposes are cover for a single purpose you will not find in any vision statement, home page, or even in a shareholder disclosure briefing. The primary purpose of every social media system is to get you to buy stuff. This stuff might be physical objects, experiences, or even some way of seeing the world (like a political or religious belief). The more completely any given system can capture your attention—become a way of life rather than a service you use to build your life—the more effective the service will be at influencing your decisions about what to buy, experience, and believe.

How does data fit into this influence operation? Put simply, the more someone knows about you, the easier it is for them to influence you. To put this in terms everyone can understand—your mom knows what you're thinking because she's seen your personality develop from the beginning. She knows that funny face you make when you're excited about something new and the way you pout when you don't get your way.

As frightening as it might be, companies that run social media systems want to know more about you than your mom has ever known.

While it's impossible to consider every possible way a company can use information to influence your decisions and beliefs, a small example will be useful. Consider the problem of an advertiser asking the question: *what kind of car advertisement will most likely make you click?* Or, perhaps, *how can I influence you to buy a new car?*

The advertiser wants to be able to place the most effective advertisement in your timeline on some social media site, in your search results, in front of that video you're about to watch, or in a thousand other places they can put an advertisement. If you think there's not much someone could learn that would influence your decision, you need to be more creative.

Have you commented on or reshared a lot of blue things? Blue dresses, blue books, blue shoes, blue bowls, blue kitchen appliances, etc.? If so, it might be that blue things attract your attention, and showing you a blue car will increase the chances you will click on the advertisement. Do you post (or like) a lot of pictures of people doing things outside? If so, you might prefer a car that says "outdoors type," something with off-road capabilities. Are you conservative? You might prefer a larger car or a car made in your country. What about your influencers—who do you pay attention to or trust? The kind of car they drive, or image they project, will likely influence your car-buying decisions. Did you just get married? Divorced? What other life changes might influence your choices?

There are far too many points to describe them all here—in fact, machine learning (sometimes called artificial intelligence) systems are used to ferret out these kinds of connections because humans cannot find them all.

Analyzing information to find everything and everyone who can influence you is critical to hooking you on products and ideas (even political and religious ideas).

Learning Not to Share

Sharing has become what the founder of Facebook called *frictionless;* you share without either thinking or—sometimes—knowing.[1] Because sharing is frictionless, it's hard not to overshare—to share unthinkingly. How can individual users learn not to overshare?

Stop Recording Everything

The group gathers around the steaming hot plate of food just placed on the table. "No, move over here . . . make a ducky face . . . now everyone smile . . . doesn't this just look great?" Someone slips off their chair, they all laugh and say, "pictures or it didn't happen." They finally settle down to eat and discover the food is cold. They motion to the waiter. "Can you take this back to the kitchen and heat it back up? Why did you bring us a cold plate of food, anyway?"

Life happens while we're taking pictures. Food gets cold. People grow old. Laughter and conversations are not shared, even if they are recorded.

The selfie has not just crept into our lives—it has altered the way we experience life to a degree unimaginable just a decade ago. Even the most out-of-the-way places are targets for selfie-taking crowds:

> Puffed, gray-tinged clouds roll over Odda, Norway, reflected in the quiet azure waters of Lake Ringedalsvatnet. More than 2,000 feet above, a hiker is perched atop Trolltunga, a cliff that juts out of the mountain. There's not another soul in sight—at least, that's what Instagram might have you believing. What photos of this iconic vista don't reveal is the long line of hikers weaving around the rocky terrain each morning, all waiting for their chance to capture their version of the Instagram-famous shot. Between 2009 and 2014, visitors to Trolltunga increased from 500 to 40,000 in what many consider a wave of social media-fueled tourism.[2]

While this might be great for the local economies, it isn't so good for the environment of these remote spots. How good is this for the person (or people) taking the selfie? Are they really experiencing a quiet moment in nature, a time to commune with the Creator? No—what they are experiencing is time standing in line, probably with people they don't even know.

---

1. Jagannathan, "Frictionless Sharing."
2. Miller, "How Instagram Is Changing Travel."

Remember this: everything you do does not need to be recorded. Every moment of your life does not need to be online. Even if you take a picture of the moment, it doesn't need to be posted right now—especially if you are far away from home. There's no reason to advertise that your house is empty, your family is alone—or pretend your life is full because you have a full timeline.

Enjoy the moment rather than taking a picture of the moment.

Use Privacy-Focused Services

Not taking pictures—staying in the moment—is a pretty small commitment in the quest to preserve a little privacy. We often choose the services we use and the stores we frequent because they are the "best" in some way; the best search results, the highest quality products, the fastest shipping, etc. Simply add "respects my privacy" to the list of things that make a service (store, etc.) the "best."

Web search is the first place to start—are you using Google or one of the other major search engines? Consider using a privacy-first search engine instead, like DuckDuckGo, Brave, or FindX.[3] You might find privacy-focused search engines aren't always as effective as the major players. They may be less effective because they don't index as many websites or are not as effective at sifting and sorting information.

But it's also because they don't return *personalized* results. If you search for the word *cub*, you might be looking for *Cub Scouts*, the *Cubs* baseball team, or perhaps even *bear cubs*. A search engine that tracks what you search for and click on will, over time, learn which of these you mean. Privacy-focused search engines generally require you to do a little more work to find what you want—you must be a little more specific, and perhaps a bit more patient, and dig a little deeper into the search results.

You don't need to stick with a single search engine, either. You can try privacy-focused search engines first, switching to larger search engines if you cannot find what you want. For instance, you could start your search on DuckDuckGo, moving to Google if you cannot find what you want.

Another service you might look at is email—is your provider reading your email and including advertisements? If the email service is free, they probably are. From a privacy perspective, switching to a provider that

---

3. One of the dangers with naming names in a book is companies change—new ones are started, old ones go out of business, business models change, etc. Take these names as examples rather than direct recommendations; it's pretty easy to find privacy-focused services in each of the areas discussed here.

charges you for email can make sense if it promises not to use or sell information gleaned about you to anyone else and doesn't include advertising as part of the package.

Even simple services, like text messaging, are a place where you can improve privacy. Text messaging services you use on your phone are not secured in any way. Some government officials in the United States have suggested inserting messages into text messages when users are deemed to be "spreading disinformation."[4] Instead of using text messages, consider using a service like Signal, which encrypts your messages end-to-end, which means even the service provider cannot read them.

There will always be a trade-off between privacy and convenience—you must decide how much of a price you're willing to pay to keep your life private.

Use a Privacy-Focused Browser

Most people only access the Internet through a browser. It only makes sense, then, that your browser leaks a lot of information about you, including the history of what you've looked at on the web, how long you spend on each page, how quickly you type, what you've searched for, etc. Many people know about cookies, but they don't know about the many other ways you can be tracked.

For instance, your browser (or your computer) can be *fingerprinted*. The bookmarks you've saved, the extensions you've installed, and other pieces of software on your computer can come close to identifying individual users (much like you can often identify a person by tracking where they drive or even where they walk in a store).[5] Another way you can be identified is by associating your phone's location with your computer or even sending inaudible audio signals between them.[6]

You should use privacy-focused browsers, like Brave, as much as possible. Some sites will not work with a privacy-focused browser, so—like search engines—you can use several. You could use Brave for most of your browsing, then move to a less private browser like Opera if you run into sites that will not open on Brave. You can use Edge or Chrome if you find something that will not work with any other browser.

Finally, you should configure your browser to clear out all personal information every time you close it, and you should close it often. Long-lived

4. Korecki and Daniels, "'Potentially a Death Sentence.'"
5. Laperdrix et al., "Browser Fingerprinting."
6. Arp et al., "Privacy Threats through Ultrasonic Side Channels on Mobile Devices."

browser sessions are bad for your privacy, your security, and your ability to focus.

Choose Small Things

We are generally impressed with big things—big churches, big political meetings, big stores, etc. Sometimes we prefer larger because it is more consistent. A big shopping site can almost always get us that new pair of pants in two days (or less). A large chain store has a consistent selection and quality no matter where we go in a country or even in the world.

However, the convenience of big things is offset by the ability of bigger companies to surveil their customers. You get the same drink (and the same quality) no matter where you go when you shop at a global coffee chain. On the other hand, you also give that coffee chain a lot of information about you, such as how often you drink coffee, where you travel (and when), how long you stay in each location, and even how many people you tend to socialize with. If the chain is smart enough, they can often figure out who you socialize with.

You can counter the constant surveillance we face by going small and going local—frequent local coffee shops instead of global chains. Find "hole-in-the-wall" restaurants where you not only enjoy the food, but you can get to know the owners and the staff at a personal level. When shopping online, favor smaller shops or buy directly from the manufacturer rather than through a large shopping intermediary.[7]

Going small and going local doesn't apply just to shopping. Build a local community, even within a larger church—live life in the flesh with physical gatherings rather than social media. The more you live a real life in a small community, the less information you will give social media companies leverage.

Keep Things Local

Another place where we need to think about trading convenience off for privacy is using cloud services to store photos and other files. It is effortless to upload every picture you take to a cloud service like Google Photos, edit them there, and then leave them there—forever. Every photo uploaded to these services contains a lot of information. The service can tell where you

---

7. Yes, this means you'll need to have a lot of accounts on a lot of different web sites, and a lot of passwords on those sites. You should counter this by using a password manager—you should, in fact, always use a good password manager.

were when the picture was taken through analysis and geotagging. Facial recognition can be used to figure out who is in the photo with you. The same is true of other kinds of files—and even old emails.

Make it a habit to delete old emails. Buy a Network Attached Storage (NAS) device, create a good backup plan, and store information locally rather than on a cloud service. Synology and Buffalo are two well-known brands of devices that can be used to create a "personal cloud," giving you a lot more control over your information.[8]

Turn Things Off

A lot of information is shared about you *passively*. Consider, for instance, the kind of information you are sharing when you install a fitness tracker. Your heart rate, breathing, and other vital signs are constantly monitored. This information might lead to the early diagnosis of some disease—or it might just lead to the fitness tracking company knowing a lot about your sleeping, eating, and other physical habits.

Even a more traditional social media app can tell a lot about you just by residing on your phone. If the app tracks your location, it can map out your daily routine, who you spend physical time with, what kinds of things you like to eat, what kinds of stores you shop in, etc. Just tracking your physical location can reveal a lot about who you are.

Remove apps you don't need from mobile devices. It's okay to only check up on your Facebook (and other social media service) news feed a couple of times from a web browser, rather than keeping Facebook with you all the time. Pay close attention to the privacy controls on your mobile devices, like phones—you can often turn off location services for apps.

What about limiting apps to getting your location when you are "using them"? Most apps don't stop running when you close them; they continue running in the background. So long as the app is running, it will have access to your location services.

Sometimes it might just be best to leave your phone at home or in your car, rather than acting like it's a permanent part of your body.

---

[8]. Once again, however, it is important to emphasize that if you buy a device of this kind and build a private cloud service, you must make certain your data is backed up. Cloud providers like Google do protect your data from being lost, but they can charge you some part of your privacy for the service.

Use a Virtual Private Network

There are two equal and opposite errors about virtual private networks (VPNs). On the one hand, some believe using a VPN will completely hide your location and allow you to use Internet-based services without worrying about privacy and security. On the other hand, some believe there are many other ways to track your activity online that using a VPN doesn't help. As with everything else in the technology world, both are partly true.

Most VPNs simply encrypt your data as it travels between your computer and other computer connected to the Internet, as shown in the illustration below.

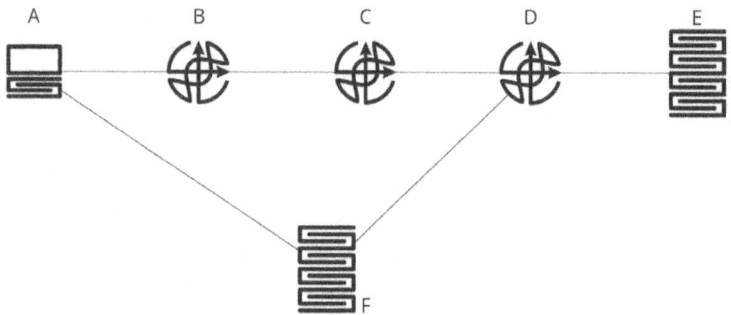

Figure 01: Data Flow in a Network

A is your computer (or cell phone, or tablet, etc.). B, C, and D are network devices of some kind (what they are and how they work are not important here). E is the physical device (server) on which the service you are trying to use runs. Without the VPN, your data passes through B, C, and D, and on to E. This data is often encrypted, so these devices cannot read the information you are sending—but sometimes it is not encrypted. The service, running on E, can "see" the entire path from itself to your local device.

When you use a VPN, your traffic is encrypted to some server attached to the Internet from your device. When your data reaches the VPN server, it is unencrypted and transmitted to its destination. In this case, traffic is encrypted from A to F. At F, the traffic is unencrypted and forwarded to E.

If this is all a VPN does, why does it matter?

First, from the server's (E's) perspective, your data originates at F rather than A. This shift in origination can (somewhat or sometimes) hide your actual location, reducing what the service knows about you. Second, any device or user that can see your data as it moves from your local computer to the VPN server cannot understand it—the data is encrypted so intermediate devices cannot read it. Without a VPN, the user sitting next to you in the corner coffee shop can intercept and potentially see the contents of what you are sending and receiving. A VPN prevents this from happening. Without a VPN, the provider you use to reach the Internet, such as the cable provider, can read what you are sending to and receiving from servers on the Internet. Using a VPN prevents the local service provider from being able to read your data.

Anonymization

Many services claim to anonymize data about you. The idea is that while you can be placed into a group, and the characteristics of the group can be determined, you, as an individual, cannot be picked out. Anonymization is one of the primary ways organizations like the International Association of Privacy Professionals (IAPP) suggest companies protect their users' privacy. If companies anonymize data to protect individual users, isn't this enough to protect your privacy?

In a word, no. A determined analyst can often recover personal identities from anonymized data sets. Xu and colleagues discovered they could identify individual users from anonymized location information publicly provided by cell phone network providers.[9] If the information has enough detail to be useful for marketing, it probably has enough information to identify who you are.

If You Have Nothing to Hide . . .

One final argument is worth considering: *if you've done nothing wrong, you have nothing to fear.* This phrase is repeated often enough, but it's hardly ever examined.

Let's begin in the most obvious place: *how many people have never done anything wrong?* I doubt the number would be anything greater than zero in any nation at any time in history. Part of the point of privacy is to allow us all to make mistakes—and learn from those mistakes—without being publicly condemned. This argument fails from the very beginning just because no one is ever perfect.

But there are not-so-obvious problems with this line of thinking, as well. In a world where public morals shift faster than the wind in a tornado, and the ire of public shame is just as bad as being caught in a tornado, everyone has said or done something that someone, somewhere, can and will

---

9. Xu et al., "Trajectory Recovery from Ash."

consider offensive. This reality doesn't mix well with living in a world where every word and deed is faithfully recorded . . . forever.

This argument also assumes the only thing anyone should fear from constant surveillance is being caught doing something wrong. However, much of the surveillance that surrounds us is focused on structuring the information and choices we are presented with to nudge us in particular directions. Not everything is about right and wrong in the world of surveillance.

Conclusion

We can, and should, break the cycle of information to influence enabled by constant surveillance. This chapter is too short to do justice to the many ways individuals can protect their privacy and reduce the amount of information they give to big social media companies.

It's essential, however, to set realistic goals. It's easy to look at the many ways companies surveil our individual lives and the convenience they give in return for the information they gather and *give up*. "I just cannot be that careful, I just cannot think of all ways these companies take information from me, and I cannot counter them all . . ." Accept that you cannot drop off the grid and lead a perfectly private life. Accept that you must often make choices between privacy and convenience.

But don't give up. You should make choices about thoughtfully selling your information for convenience, rather than just "going with the flow," accepting frictionless sharing as a part of life you cannot change. You cannot block surveillance entirely, but you can reduce its effectiveness, and you can draw lines at which you say "this far and no farther."

7

Refuse to Perform

"All the world's a stage, and all the men and women merely players." When Shakespeare put these words in the mouth of Jaques, he meant to shock through the comparison, to show the audience they, too, are actors on a stage, being watched and playing a part. Each person is both actor and audience in the drama of life; the only difference is the size of the audience. Actors in a play live on the stage or the platform, and hence can have a large audience, seen by all, and having no real contact with their audience. Individual life happens in view of a much smaller audience and intimate contact with the audience.

The great playwright recognized the flattening effect of the audience on the performer, how the audience always desires to place the performer (or at least the character being performed) into a simple-to-understand shadow of a real person.

What happens when everyone has a stage on which to act and—at the same time—the ability to interact with every member of their audience? Social media.

What can the actor do when they find the acting itself is making them sick? They can refuse to perform. What can the audience do when they discover the play has consumed their lives? They can refuse to pay attention—they can refuse to perform the part of the audience.

Refusing to perform is the second of the three ways individuals can counter the adverse effects of social media. Refusing to perform begins by rejecting the highlight reel.

After considering the importance of rejecting the highlight reel, this chapter will look at three actions every user can take to help them reject the highlight reel and use social media in a healthier way. It is much easier to say you should reject the highlight reel than do it.

Mental Resistance

The first line of defense against the negative effects of social media is developing good mental habits—healthy ways of thinking about how you use social media. Three specific areas to address are rejecting the highlight reel, ignoring likes, and disregarding trolls.

Reject the Highlight Reel

Comparison, according to Ray Cummings, is the thief of joy.[1] And yet, we compare ourselves to others almost constantly.[2] This constant comparison is especially true in the virtual world of social media.

Consider what occupies just about every social media feed. If you look at a Facebook profile, what are you most likely to see? A steady stream of perfect pictures of family and friends, interspersed with almost breathless reporting on the most recent award or adventure. If you look at just about anyone's Instagram feed, what are you likely to see? A steady stream of inspirational (or funny) quotes, perfect meals served on perfectly set tables with perfect friends, gorgeous scenery, or gorgeous bodies decorated with the most perfect makeup and clothes imaginable. What about TikTok? Hilarious videos of the perfect prank or "fail," or perfect dances performed by people with perfect bodies. LinkedIn? A constant stream of new and important professional information is interspersed with a rising tide of accomplishments and awards. You will also find a few quotes to inspire you to work harder or better or live a better life, mixed in for good measure.

If you spent all day doomscrolling through everyone else's feeds on any social media site, it would be easy to believe everyone except you leads a practically perfect life. No one else has setbacks. No one else is ever depressed. Everyone else is out with friends on daring adventures and leading interesting, successful lives.

All of this is a lie.

---

1. While this quote has been attributed to Theodore Roosevelt, C. S. Lewis, and others, the first place I can find it is in Cummings, *Are You Following Jesus or Just Fooling Around?*, 81.

2. About 10 percent of our thoughts are spent comparing ourselves to others, according to Summerville and Roese, "Dare to Compare."

Well, maybe not all of it—but most of it is. What you see in all these social feeds is the highlight reels of the lives of others. Social media users do not even need to be aware they are performing—the resonance of the media draws each user in, so the line between just living our lives in front of those we care about and living our lives on a stage is blurred beyond recognition. Social media is nonymous rather than anonymous—a point made earlier in this book. What happens online does not stay online.

Amy Summerville argues these self-comparisons are harmful because you tend to compare yourself to those you perceive are better than you in some way—a better cook, a more social life, more beautiful persons, more financially successful persons, etc. We consistently compare our (necessarily) average skills or success to the skills and success of those who are the best in any given area.[3]

Some experts argue it's better to compare yourself to others you believe are equal or worse than you. Summerville says, "Social cognitive psychologists have long known that when we want to feel better about ourselves, we make comparisons to people worse off than we are (or think of ways that things might have been worse than they are)."[4]

Other researchers, however, argue that any kind of comparison at all has adverse effects on your mood and perceived self-worth. For instance, Steers and company say, "any benefits gained from making social comparisons may be temporary whereas engaging in frequent social comparisons of any kind may be linked to lower well-being."[5] They back this assertion up with a small but strongly suggestive study of feelings of depression tied to *any* use of Facebook.

The idea that any comparison with others has negative repercussions does not originate with Ray Cummings (or Theodore Roosevelt, or any other famous figure). This idea can be found in the Scriptures. In 2 Corinthians 10:12, Paul says, "But when they measure themselves by one another and compare themselves with one another, they are without understanding."

A healthier attitude is the one Paul expresses in Philippians 4:11, "I have learned in whatever situation I am to be content." As they say on the Appalachian Trail, *hike your own trail*. It's okay to watch others learn; it's not okay to watch others compare yourself to them.

---

3. Summerville, "Is Comparison Really the Thief of Joy?"
4. Summerville, "Is Comparison Really the Thief of Joy?"
5. Steers, Wickham, and Acitelli, "Seeing Everyone Else's Highlight Reels," 704.

Ignore the Likes and the Trolls

Human beings are drawn to numbers like moths are to a flame—numbers, especially when they can be used to tell us who we are, are irresistible bait for our self-image.

> Quantification becomes the way we evaluate whether our desire for more is being fulfilled. If our numbers are rising, our desire is met; if not, it remains unmet. Personal worth becomes synonymous with quantity. . . . We want to "win" the confidence of our friends, to accumulate a capital of "social connections, honourability and respectability" that can be exchanged later within our social system. . . . when combined with metric visibility in Facebook, our desire for social capital leads us to internalize the need to excel in a quantified manner.[6]

We believe measurements tell us more about ourselves than our friends or family (or even the Scriptures) because numbers don't or can't lie. Further, if something can be measured, we want to move the numbers in the right direction all the time.

Rather than being healthy, we want to be perfectly fit. To be healthy is to rise to an objective standard, but fitness is a never-ending quest—you can always be "more fit."[7] We would rather have a limitless supply of "friends" on a social media site than a few close friends in real life because more friends are always better.

The key to stopping this quest for ever greater numbers is to ignore the numbers. Refuse to quantify your relationships in this way. Almost every social media network puts the number of likes, reshares, and other numbers directly in front of you—this is an intentional part of the experience to get you emotionally hooked. While it's very hard to ignore the numbers, it's healthier if you can.

Trolls are the other side of the likes—people who counter each positive post with a put-down. Trolls live to start arguments, drawing attention to themselves. It doesn't even matter if they "win" the argument they start, so long as they can "win" the emotional way by drawing more attention to themselves than you get from the original post.

Ignore trolls.

---

6. Grosser, "What Do Metrics Want?," 2.
7. Anticoli and Basaldella, "Shut up and Run," 1553.

Practical Resistance

Rejecting the highlight reel, ignoring likes, and disregarding trolls are mental habits of resistance to the power of social media—how can you implement them at a practical level? By controlling your eyes. Customers in the world of social media are called eyeballs because grabbing and keeping your attention is the metric of success.

Controlling your eyes helps you stop performing by helping you break the social media habit and untying your worth from the feedback you receive in the virtual world. Further, it breaks the virtuous cycle social media networks rely on to drive pervasive surveillance. Finally, performances require an audience—if you decide not to be a part of the audience, you are helping others stop performing.

Minimize Notifications

How many times a day do you receive a notification from a social media network on your device? How hard is it to resist looking, even when you're in an important meeting or the same physical space with someone important?

How often do you think a social media network could find a reason to notify you? If every user on a social media network with 500 million users posts one new update each day, the service has 500 million updates, each of which could be used as an excuse to notify you. The service can notify you at any time, any place, about anything.

The ability to notify you anytime about anything plays directly into the operant conditioning playbook—semi-random reinforcements scheduled across time, building a habitual reaction to some external stimulus. Each time you respond to a notification by looking, you deepen your habit and strengthen your emotional connection to the social media service.

A simple suggestion: turn off all notifications from social media networks on all your mobile devices. No more dings, rings, tones, or buzzes from your phone vibrating on the table. Reducing the notifications your devices send might even help reduce the number of times you think you're being notified—even when you're not.

Go further than this by turning off the "live badges" on your mobile device. Do you need to know how many new Facebook messages you have every time you pick up your phone? Is your fear of missing out on something happening in the virtual world stronger than your desire to fully be with someone in the real world?

Notifications and badges are ways social media draws you in from the real world to the virtual one. Refuse to be drawn in.

Choose Intentional Interaction

When you're brave enough, you can (and should) remove all the social media apps from your mobile devices. Use a web browser to interact with social media services instead. Removing these apps allows you to choose when you are going to interact with the social media network. You can decide how many hours each day you will spend on Facebook and when you will spend that time. Rather than being pulled into paying attention, you can choose when to pay attention.

A nice side effect of taking all the social media apps off your phone is they can no longer track you. Social media apps are not only notifying you of their presence when there's new activity; they are using information from your mobile device as a form of surveillance. When you install an app, the app gains access to your emails, texts, contacts, and location—at least. Do they need to know all of this?

Location information is particularly rich, especially when combined with the location of other users. The speed at which you're traveling indicates how you are traveling, and previous trips can be used to predict where you are headed right now. If your location data shows you regularly stopping by one brand of coffee shop, then you are likely to stop at that same brand no matter where you are. If you often end up in the same location as a small group of people, then you are likely friends with that group—or you "should be."

If you want the ultimate control over your social media usage, remove as many social media apps from your devices—your phone in particular—as you can stand. It's better not to have social media apps installed on any device at all. If you must install them, install them on the least portable device you own or the one you take with you least often.

Take Social Media Sabbaticals

Sabbaticals are built into much of the rest of our lives—even if we don't take them. For instance, in the Jewish and Christian worlds, there are weekly sabbaticals. There are sabbaticals related to holidays, and some companies (particularly in the educational field) allow their employees to take sabbaticals. There is, in fact, an "enforced sabbatical" in everyone's life every night in the form of sleep.

Sabbaticals are often recommended from social media, but we tend to take them as seriously as we do the other sabbaticals in our lives. What if we miss something important while we're on a sabbatical from social media? For those who write, teach, or otherwise rely on their online footprint for at least some part of their livelihood, there is a real concern about losing your audience.

Several points counter these concerns.

First, sabbaticals do not need to be extended to be effective. Taking several shorter sabbaticals each day can be effective at creating mental and emotional distance between the online and offline worlds. Perhaps a longer sabbatical once a week—just don't check any social media on Sundays—can be effective.

Second, sabbaticals can be built into the flow of everyday life. If you set aside time to work on projects requiring deeper thinking or that must be handled "interruption free," you are already preventing yourself from constantly engaging with social media. Sabbaticals of this kind, then, can be built into your schedule.

Finally, you will find that your concerns about losing your audience or missing something important will fade as you become more focused on the here and now and less on the flow of information coursing through the online virtual world.

Sabbaticals need to be adjusted to fit into your life and the way you live—but they can be an effective tool to counter the negative effects of social media.

Don't Filter

While everyone likes to be beautiful on camera, it is easy to cross a line between wanting to look your best and being liked because of how you look. Crossing this line can have significant repercussions in real life. Perhaps the most extreme example is the rising use of cosmetic surgery to look like a filtered photograph of yourself. Doctors are reporting instances of *Snapchat dysmorphia,* where patients ask for surgery to look like a filtered selfie of themselves—larger, fuller lips, thinner cheeks, removal of bags, etc.[8]

Perhaps a good rule of thumb is never to use anything that reshapes your face or body as a filter when posting an image. Softening, changing lighting, and adding cute ears are okay, but reshaping your face to look different than you do in real life invites a disconnect that can eventually seep into your self-identity and cause major psychological distress.

8. Hosie, "More people want surgery to look like a filtered version of themselves."

This rule of thumb can be extended to the rest of the images you present online. In the modern world, you cannot simply not have a public presence online. Even if you don't have a Facebook account, your friends do—and any photos with you in them will be tagged.

Given you will have an online presence, you should take control of that presence, but in a way that honors the truth. Don't exaggerate accomplishments, and don't post every little accomplishment to social media. Refusing to brag in this way can be particularly difficult in dating (finding a mate) and professional life. Employers and potential spouses are always looking for "the best available option," and "everyone tells little white lies to make themselves look better on paper."

Spinning things to enhance your accomplishments is borderline, but lying or exaggerating is simply unethical and can lead to nothing other than disappointment in the long term. Tell the truth about your successes and failures (and don't exaggerate your failures, either).

Another way to approach the filtering problem is to focus on ideas and events rather than people. Anything you might classify as gossip should be avoided. Focusing on ideas and events can help control the urge to perform.

Concluding Thoughts on Resistance

The first step you can take to resist the harms of social media is to simply *stop giving social media access to the external and internal triggers that ground habit formation.* Stephen Wendel describes user experience designers building an *action funnel* containing five mental events: "a *cue,* which starts an automatic, intuitive *reaction,* potentially bubbling up into a conscious *evaluation* of costs and benefits, the *ability* to act, and the *right timing* for action."[9] The key to not paying attention is to break the automatic movement from *cue* to reaction without evaluating costs and benefits.

---

9. Wendel, *Designing for Behavior Change,* loc. 1400.

8

(Re)Learn to Think

While the previous chapters have focused on the impact of social media on our view of the person and the results of flattening the person, this chapter will focus on the flattening of information, or rather the impact of social media (and related systems) on the way we think. Thinking is a learned skill, just like riding a bicycle, driving a car, or painting. The way we think reflects—and is a reaction to—the way information is presented, how we have successfully navigated the world, and what strategies we use for finding truth.

The adverse effects and suggested counters considered in this chapter apply to services beyond social media, such as search engines and shopping services.[1] Continuous partial attention and the move from flow to factoid will be the two focal points of this chapter.

Continuous Partial Attention

People—especially people born and raised in a world full of screens—believe they are good at multitasking. They believe they possess the ability to answer a private (or direct) message, doomscroll through a social media feed, watch short (funny cat) videos, or even play a game while talking to

---

1. Many of these services, however, can be considered social media because of their strong social elements. Retail services rely on gathering information about you to shape how and what to offer and at what price—each user can be offered the same product at a different price, or substitute products can be offered. Search engines, likewise, can (and do) offer different results to each user.

others or watching a movie. Here's a short counter to this belief: Humans stink at multitasking.

Humans can't multitask at all. Instead of multitasking, the human brain just switches from task to task *really* fast—so fast it seems like it's possible to do both things at once. When you're multitasking, however, you're not doing either task as well as you can. This quick task switching might be okay when you do two simple tasks, like making cookies and following a conversation. However, the more complex the two tasks become, the more likely you are to make a mistake. There are solid reasons why texting while driving is never a good idea.

We normally multitask to improve our productivity—why spend time just listening to this presentation when we could be reading and deleting some emails at the same time? When multitasking crosses over into a constant state, motivated by a fear of missing something important (fear of missing out), it becomes *continuous partial attention*. Linda Stone describes continuous partial attention this way:

> Continuous partial attention involves a kind of vigilance that is not characteristic of multitasking. With CPA, we feel most alive when we're connected, plugged in, and in the know. We constantly SCAN for opportunities—activities or people—in any given moment. With every opportunity we ask, "What can I gain here?"[2]

When you are scrolling through the feed on a social media network while having a conversation with a person in real life, you are probably engaging in continuous partial attention. You are not just concerned with increasing your productivity but making sure you don't miss out on something that happens in the virtual world while still participating in the real world.

Continuous partial attention isn't just rude—the focus on missing out creates a constant sense of urgency, sometimes bordering on panic. While you are in this state of crisis, your body's fight-or-flight response is triggered, just like it would be if some wild animal were attacking you. Continuous partial attention leaves you physically exhausted, overwhelmed, and unfulfilled.

It is *not* a crisis to miss something on social media. That email can and will wait for another hour—even two.

The best way to break the continuous partial attention habit is to take control of your attention using the same techniques and tools described in

---

2. Stone, "Beyond Simple Multi-Tasking."

the last chapter. Reduce notifications, turn off badges, remove apps, and make your interactions with social media services intentional.

Shifting Perception of Truth

It's widely accepted that long-form reading is a dying form of communication. According to Jean Twenge and her colleagues, the average eighteen- to nineteen-year-old spent about two hours per day on social media in 2016—a number that is hard to measure (because of continuous partial attention) and is likely rapidly increasing. Online gaming is making similar strides in consuming time. At the same time, the number of people in the same age group who did not read *any* books for pleasure has increased from around 10 percent in 1976 to about 35 percent in 2016.

Anecdotal evidence points in the same direction in a different way. Books of around 90,000 words were once common; today, most popular books are less than half this length. Blogging began as a relatively long form of writing; posts ran to a thousand or more words. The advice given to bloggers today is to keep their posts under five hundred words, with the ideal length being less than 250 words.

In the words of Caitlin Flanagan, "Twitter did something that I would not have thought possible: It stole reading from me. What is it stealing from you?"[3]

What is the impact of this shift from long-form reading to video, social media feeds, and other ways of interacting with information?

Immediacy

Those who forget the past are doomed to repeat it. All too often, those who do remember the past are doomed to watch those who forget the past repeat it. These are trite sayings we've all heard thousands of times, but the truth they contain isn't going to wear out by being repeated. Understanding the past helps us understand human nature, what is possible, what is not possible—and even what's been tried and failed.

How many times can humanity try some form of extreme government-enforced collectivism, such as communism and socialism, and see it fail before we decide there just isn't a bucket of gold at the end of that particular "rainbow"? Those who know history know "it's never really been

---

3. Flanagan, "You Really Need to Quit Twitter."

tried before" is a terrible excuse for risking millions of human lives in the attempt to "get it right this time."

Social media, on the other hand, is immediate—what's happening right now. There's little concern for what happened a hundred years ago, last year, or even (all too often) the day before yesterday—except as fodder for factoids or memes—because there just isn't time to pay attention. When you live in a world where there's always a new funny or cute cat to watch, what is the past worth? Not very much.

This immediacy creeps into our lives. There is no moment but this moment, every moment. There is no past, there is no future, only an eternal now, an everlasting dull moment filled with chatter about which we don't care. In our desperation to find meaning, we pay attention to everything all the time—continuous partial attention.

This immediacy also creeps into our thinking, often taking the form of what C. S. Lewis calls "chronological snobbery."[4] The past is strange, filled with less enlightened people who didn't have "science," who held to beliefs and ideals we consider worse than untrue. These people are all evil, their statues and works destroyed, their lives erased from existence.

This single-minded focus on the immediate destroys the narrative—the very fabric—of our lives, communities, and culture.

Factoids

Computers and books structure information differently. Neither facts nor ideas stand alone in a book; they are always connected to a flow of thought in the form of an argument or a narrative. Information is not a commodity in long-form writing but is rather offered in support of some greater objective.

On the other hand, computers structure data (really just the facts) so it can be searched. Information technology relies on information theory, which converts all communications into transmissions along a channel and gauges the outcome of communications by its impact on the world in terms of its "surprisal effect."

What is the impact of this shift from the flow of an argument to a narrative to the factoid? Facts that stand alone—without context—can be made to say anything. As the old saying goes, a statistician is someone who can put their head on a hot stove, their feet in a bucket of ice water, and say, "on average, I feel fine." Replace statisticians with data scientists, and you have pretty much described much of what is wrong with big data.

4. Lewis, *Surprised by Joy*, 206.

Since computers primarily organize information for searching and access information through computers, we are becoming habituated to looking at the world through factoids. Nicolas Carr says:

> We seem to have arrived . . . at an important juncture in our intellectual and cultural history, a moment of transition between two very different modes of thinking. What we're trading away in return for the riches of the Net—and only a curmudgeon would refuse to see the riches—is what Karp calls "our old linear thought process." Calm, focused, undistracted, the linear mind is being pushed aside by a new kind of mind that wants and needs to take in and dole out information in short, disjointed, often overlapping bursts—the faster, the better.[5]

We have come so far down the road of dividing facts from their context that we no longer seem to consider narratives as containing anything more than inferences. While facts are true regardless of their context, anything with a narrative element—worldviews, flows of argument, historical events, etc.—can be changed to fit the desired outcome.

This focus on the facts alone represents a radical shift in the basic thinking process we use to arrive at what is true and real—including the way we relate to the very Word of God—a topic addressed in greater depth in the next chapter.

Rejecting the Shift

An excellent place to start in rejecting these shifts in thinking is unhooking from social media. The previous chapter provided some ideas towards this end. Unhooking from social media, however, is not enough. We must replace the bad habits of thinking promoted by social media with better habits. Doing so will require some measure of self-discipline, hard (mental) work, and being willing to follow the evidence wherever it might lead. These are difficult for the modern mind, attuned to convenience, to accept—but they are the only answer to the problem.

Begin by reading. Cultivating a habit of reading tens of books a year (even most readers in our modern culture read one or two books a year) can help "de-factoid" your thinking. What kinds of things should you read?

You should read books you agree with, of course. You should also read books you disagree with—especially if they make a strong case for an idea you don't agree with. You should read short books, of course, but you should

---

5. Carr, *The Shallows*, 10.

also read long books. Read books that are difficult to read, even if it's hard work.

C. S. Lewis's advice of reading old books is good advice, as well. If you read old books, you will encounter old ideas—this can help you overcome the chronological snobbery so prevalent in modern culture. Lewis says:

> In the first place he made short work of what I have called my "chronological snobbery," the uncritical acceptance of the intellectual climate common to our own age and the assumption that whatever has gone out of date is on that account discredited. You must find why it went out of date. Was it ever refuted (and if so by whom, where, and how conclusively) or did it merely die away as fashions do? If the latter, this tells us nothing about its truth or falsehood.[6]

Reading is not—cannot be—the only solution to this problem, however.

A second way to reject the shift social media causes in our thinking is to intentionally contextualize every fact you encounter. If someone presents a fact to support some argument, don't accept the fact at face value. Instead, chase the fact down to its source and context. A simple search through your favorite search engine—and accepting the first search result that supports or denies the fact—is not enough.

When listening to an argument—any argument—contextualize the facts presented by outlining the argument being made. Where does this fact fit into the argument? Does the fact support the argument being made? Is the fact true?

Think about the argument, as well; don't limit yourself to the facts. Is the argument logical? Are there contradictions within the argument itself? What are the logical consequences if you accept this argument as true? What are they if you reject the argument as false?

The bottom line is your best defense against the shifts in thinking is better thinking. Better thinking isn't something you will get just by reading a book, listening to a podcast, or an online video series. It's something that is going to take a lifetime of learning.

There's no better time to start than now.

---

6. Lewis, *Surprised by Joy*, 206–7.

9

Christian Witness in a Social Media World

CHRISTIANITY HAS NOT PLAYED a prominent role in this book so far, other than the general Judeo-Christian view of the human (anthropology). This chapter, however, will mark a change by speaking directly to Christians about living as a Christian in a social media world. What does a Christian witness look like in a modern culture tuned to social media? What does Christian apologetics look like in a social media world?

Relationships

"Let me tell you about an experience I had that shows Christians can win the argument." The apologist was in front of a packed house, riffing on experiences and countering questions from the audience about the effectiveness of Christian thought in the modern world. "I once had an atheist stand up in the middle of my lecture and argue he couldn't believe a good God would create a world with so much evil in it . . ." The audience was on the edge of their seats—everyone sitting in that room had probably been confronted with some version of the problem of evil.

"Well, before I could even answer," the apologist continued, "another man stood up in the back of the room. He explained that he was a pastor who had brought thousands of people to Christ all over the world. However, he was also an abortion survivor. Certainly, if God can bring good out of the life of a man who survived an attempted abortion, he can bring good out of whatever evil you might see in this world! Let me tell you, that atheist shut up and walked out of the room!"

The audience clapped and whooped—Christians *can* win arguments like this!

Ignoring the shallow response to the problem of evil (any well-educated atheist with sharpened debating skills could shred the answer), what was the result of the exchange?

The atheist shut up and left the room.

Shutting people up is the last thing we need in a world infused with social media—a world where relationships and people have been flattened and quantified, their commercial value drawn out.

People don't care about what you know, who you are, or what you think until they know you care about them.

Relationships in a social media world are thin and unsatisfying. The young, the beautiful, and those with interesting "skills" are well-liked, but those likes are subject to immediate cancellation any time you say something falling outside the bounds of acceptable discourse.

During the coronavirus pandemic, people were celebrated for not trusting the vaccine—but when the political winds shifted, those same people were celebrated for pushing people to be vaccinated. Those without the political clout required to make the shift were celebrated one moment and canceled the next. Regardless of the effectiveness of the vaccines or the wisdom of any policy related to them, the shift in public opinion—and the lives ruined on either side of that shift—illustrate the tenuous nature of popularity in a social media world.

Genuine friendship, lasting through thick and thin, through popularity gained and popularity lost, is a true treasure in the modern world. If you want someone to listen to you, treat them like a real person, rather than the flattened version of themselves they present on social media.

There is theological warrant to put the relationships first. Paul says, "For by works of the law no human being will be justified in his sight since through the law comes knowledge of sin," in Romans 3:20. If the law, no matter how closely followed, cannot save, what can save? Faith in Christ. What is faith in Christ? It's not just knowledge but a *relationship*.

Begin by building relationships rather than winning arguments.

The True Myth

The modern world is obsessed with clearly dividing narratives from facts. Anything told with some element of time, whether a story or an argument with elements, is suspicious. The postmodern impulse has given way to broadly accepted critical theory. All narratives are necessarily told from some perspective to achieve a specific goal.

In critical economic theory, narratives are told to morally justify the divide between the wealthy and the poor, keeping the owner in power over the laborer. In critical race theory, narratives are told to justify one race gaining and keeping power over another. In every case, narratives are a way to gain power rather than a way to describe truth. The focus on scientific (and medical) facts has sharpened the sword that divides narrative from truth.

This attitude towards narrative has a strong foothold throughout the Christian world, as well. For example, Marshall Wicks, a well-known theologian, speaking about the relationship between God and time, says:

> At the heart of the theological differences between the Christian eternalist and temporalist is a different estimate of what constitutes such a good reason as not to take some scriptural representation of God literally. Opponents of divine timelessness want the text to be taken literally unless there is just cause to relegate it to anthropomorphism. While this initially seems like a responsible thing to do, it does place a heavy burden on the interpreter's ability to decide just cause. A more responsible standard may be that one should be completely open as to how a text is accepted. . . . Narrative texts in particular are notorious for anthropomorphizing God's interactions with man.[1]

Whatever position you might hold about divine timelessness, dismissing scriptural narratives as somehow containing "a different kind of truth," or perhaps even "less truth" than passages containing "direct teaching" (called *didactic passages)*, is dangerous—especially in a world that already dismisses all narrative as ideologically tainted.

The natural reaction is to abandon the narratives of Scriptures and just argue facts. If the world wants to talk in short quips and argue with factoids, Christians should adapt themselves to this way of discussion. Allowing yourself to be taken to the culture's field, playing by the culture's rules, will never work.

This moment in time calls for something else. Instead of abandoning the narrative to fit into the larger culture, Christians need to work counter to the culture by emphasizing the truth narratives contain. Christians need to argue that narratives contain the same kind of truth every scientific experiment, medical proclamation, and social experiment contains. While some

---

1. Wicks, "Is It Time to Change?," 50–51.

truths might be better presented in a narrative, there is only one kind of truth.

C. S. Lewis's conversion to the Christian faith is a perfect example of infusing an old narrative with new meaning that leads directly to God. While Lewis was raised in a Christian house and culture and was very curious about Christianity, he had a problem accepting the scriptural narratives as anything more than a myth. Lewis understood how a world might need salvation, but he struggled with how a person who lived some two thousand years ago dying and coming back to life might provide that salvation. To Lewis, the narratives explaining this seemed to be like any other myth of a god who dies and comes back to life.

For Lewis, a myth moves you emotionally and mentally but is not necessarily related to truth. The gods of Babylon died and came back to life, and we know those stories are not true. Lewis continues, however:

> Now the story of Christ is simply a true myth: a myth working on us in the same way as the others, but with this tremendous difference that it really happened: and one must be content to accept it in the same way, remembering that it is God's myth where the others are men's myths: i.e. the Pagan stories are God expressing Himself through the minds of poets, using such images as He found there, while Christianity is God expressing Himself through what we call "real things." Therefore it is true, not in the sense of being a "description" of God (that no finite mind could take in) but in the sense of being the way in which God chooses to (or can) appear to our faculties.[2]

Once Lewis saw that myths could be true (or contain truth), he suddenly realized how the narratives of Jeshua dying and coming back to life could give the kind of salvation the world required. The key, for Lewis, was connecting "truth" to "myth." Some myths are true.

The world is like Lewis before his conversion—convinced all narratives are myths—only more cynical. All narratives are myths told to gain and hold power, and hence they do not—they cannot—contain truth. Christians need to counter this.

This book is too short to explore building a Christian worldview and culture that effectively counters the widespread belief that all narratives are "power myths," so some suggestions will need to suffice for now.

First, Christians need to live as if the narratives found in the Scriptures are true. Living as if the Scriptures are true is the strongest witness to their truth we can give to the world. How this works out in culture is a point for

---

2. Lewis, *Collected Letters*, 1:977.

further discussion—beyond the scope of this book—but part of living as if the narratives are true is refusing to reduce our relationship with a God to a set of facts or a set of emotions. Our modern culture is accustomed to flattened, shallow relationships because of social media. Our relationship with God is anything but flat.

Second, Christians should refuse to give up the whole of the Scriptures as the Word of God. Christians need to stop ceding ground on the truthfulness of the Scriptures just because we find the narratives embarrassing in some way. Throughout the Scriptures, God reveals himself in acts, and those acts result in narratives. Meaning, in the Scriptures, is often embedded in a stream of chronological events with complex plot lines unfolding over time. Theologians, pastors, and Christian scholars, in particular, need to stop treating the narrative passages of the Scriptures as "second-class citizens" that can only infer truth rather than directly teaching or containing truth.

Facts

This emphasis on the true narrative necessarily includes learning to defend the Christian faith from facts and logic. While arguing from factoids may not have much of an impact in a world focused on "scientific truth," facts can be used to support the truth of narratives, ultimately showing the scriptural narratives contain truth. Evolution is a helpful example.

Two things are true about evolution: at a popular level, almost everyone believes evolution gives a true account of the origins of life on Earth, and evolution's ability to explain the origin and diversity of life as we know it has failed. To be clear: evolution happens in the real world. Scientists can show that living things evolve in various ways. On the other hand, evolution simply cannot account for the origin of life on Earth, nor its diversity.

The widespread belief in evolution as a "theory of everything" is a testament to the power of narrative in our culture. It doesn't matter that evolution has failed as an explanation for the origin or diversity of life. All that matters is that the narrative of evolution has "caught hold" of the average person's imagination.

Facts, of course, are one way to argue against the evolutionary narrative—but in a world awash in facts, this path might not be very effective. Instead, Christians must show that the existing evolutionary narrative cannot provide a full explanation of reality, leads to outcomes no one would rightly accept, and propose a counternarrative.

For instance, evolution simply cannot explain irreducible complexity—and the more we discover about the nature of life, the greater the problem of irreducible complexity becomes. Socially, evolution was associated with eugenics and other theories now considered unacceptable for a reason. Have those associations been fully explored and proven incorrect, or have they been ignored for the sake of advancing the narrative?

Christians arguing on facts should never forget *the point is not to win an argument.*

Taking the Roof Off

One of Francis Schaeffer's key insights is that everyone lives according to some narrative (or narratives). However, each of these narratives has some weak point, a place where there is an obvious contradiction with reality or where the narrative fatally contradicts itself. Narratives with these kinds of faults are unlivable. To resolve this problem, the person who follows a narrative will build a "roof" to cover the point where the narrative they live by fails.

Schaeffer argues Christians must find this point—find the place where the narrative doesn't make sense—and take the roof off. Expose the self-contradiction, the contradiction with reality—whatever the weakness is—allowing the real world itself to do the work.[3]

Finding these weaknesses and taking the roof off is not easy work, especially in a world that holds all narratives exist to get and hold power. There are large metanarratives, such as critical theory, that you can learn, getting to the point where you understand the weakness in the whole system.

However, in a relativistic world, everyone takes these larger metanarratives and shapes them to fit their own experience. In the modern mind, everyone has their own truth, and that truth is expressed as a narrative to which no one else has access. To take the roof off, then, you must not only understand the larger metanarrative and its weaknesses, but you must also understand how the person in front of you has modified that narrative to fit into their life story.

Taking the roof off is, itself, a delicate process, especially in a world where challenging someone's view of themselves is often seen as a form of bigotry or hatred. This process can only take place within a relationship. Only within the context of a relationship can you understand another person's personal narrative and challenge it. Only when someone sees the relationship will not end because of disagreements will they be open to discussion that might lead to a change of narrative.

3. Schaeffer, *The God Who Is There*, 155.

10

Unfriending Dystopia

THE CREATORS OF SOCIAL media truly believed they were building something that would result in positive cultural change. People would be more connected to one another and have more open and honest discussions. Tolerance and economic activity would thrive through these systems. When you start with false premises—humans can create an increasingly perfect world, and good intentions insulate people with power from doing evil—all your utopias turn to dystopias. Modern social media is a perfect example of an unintended dystopia, where good intentions have paved the way to a virtual netherworld.

Unintended Dystopia

The virtual world created by social media is rife with ills impacting individual users and culture.

Filter bubbles are common. Filter bubbles, according to Eli Pariser, "create a unique universe of information for each of us . . . which fundamentally alters the way we encounter ideas and information."[1] It is easy to live your entire life hearing only things you agree with, particularly if you agree with the dominant narrative—there is no conversation but rather an enforced orthodoxy.

The all-too-common result of the dominant progressive narrative controlling the levers of social media filtering is the chilling of speech through the spiral of silence. Social experiments have shown individuals will

---

1. Pariser, *The Filter Bubble*, Introduction.

self-censor in the face of public disagreement.[2] Elisabeth Noelle-Neumann argues this self-censorship occurs because of fear of isolation (or fear of missing out).[3]

The result is widespread isolation leading to social harms, including lower life satisfaction,[4] lowered satisfaction in intimate relationships,[5] depression,[6] and increased isolation.[7] Some studies have even tied consistent social media use to a higher chance of divorce[8] and an increase in jealousy within marriages.[9] Others appear to show that divorce is in part a social phenomenon; marital dissolution can be tied to a user's social media connections divorcing.[10]

The effects of social media on its own and driving people to attach themselves to screens all the time can hardly be overstated.

Should We Just Disconnect?

How can you counter these adverse effects? How can communities counter these negative effects? The most straightforward response is to unplug. Stop using social media entirely. Stop using screens. Live in the real world and ignore the virtual world of social media.

Will this work? No.

Social media provides a great way to connect to people. Social media is not good if you use it as the primary way to connect to people. It isn't a good place to develop and maintain a fully orbed relationship. It is, however, a good way to augment relationships that already exist in the real world.

Social media provides a good way to get a message out. Blogs, podcasts, and newsletters were once the primary way you could communicate with public followers as a person or organization. Many of these methods now rely on social media as a funnel to draw new followers into your content and let existing followers know about new content.

---

2. See, for instance Hayes, Glynn, and Shanahan, "Validating the Willingness to Self-Censor Scale"; and Penney, "Chilling Effects."
3. Donsbach, Salmon, and Tsfati, eds., *The Spiral of Silence*, 6.
4. Rotondi, Stanca, and Tomasuolo, "Connecting Alone."
5. Hammond and Chou, "Using Facebook," 41.
6. Boers et al., "Association of Screen Time and Depression in Adolescence."
7. Primack et al., "Social Media Use and Perceived Social Isolation."
8. Valenzuela, Halpern, and Katz, "Social Network Sites, Marriage Well-Being, and Divorce."
9. Farrugia, "Facebook and Relationships."
10. McDermott, Fowler, and Christakis, "Breaking Up Is Hard to Do."

There are lost people on social media, too. While social media is not a good place to witness *or argue* (please don't engage in extended arguments on social media sites, it's counterproductive), it is a good place to make initial contact with people who are seeking the truth.

Unfriending Dystopia

If disconnecting isn't the best path forward, what can we do?

Reject the Californian Ideology. People are not manipulable objects. Individuals are made in the image of God, meant to be treated with dignity. Reject the shallow vision of progress that declares one set of ideas the "right side of history." Reject naturalism and its view of humanity as just another part of nature. These counters are all best accomplished through communities, and hence are discussed in the chapter on how communities can counter the effects of social media.

Refuse to perform. Don't tie your personal worth to your online popularity. Don't build your identity on an identity that attracts attention online. Don't record your life and put it online. Live your life in real life rather than online.

Maintain a defensive privacy stance. Learn to stop oversharing. Pay attention to protecting your privacy, even if you don't think you have anything to hide.

Counter social media thinking habits. Read broadly and deeply. Outline arguments to understand them fully, rather than just accepting them on their face. Research facts when they are presented—preferably all the way to their source. Treat narratives with respect, accepting that they contain the same kind of truth as facts and factoids.

These things are not going to be easy. They will take work, and they will mean giving up some convenience and potentially even some opportunities. But if we fail to work against these effects, we could lose the biggest opportunity of all—the chance to fully become the humans God intended us to be.

Appendix
Further Reading

THIS IS NOT, BY any measure, a complete list of books on the topics considered in this book. For those who would like to investigate further, however, it is a good place to start. This is an annotated bibliography, in that each entry comes with a short description. Readers should also feel free to peruse the standard bibliography for more reading ideas.

Carr, Nicholas. *The Shallows: What the Internet Is Doing to Our Brains.* Kindle ed. New York: Norton, 2011.

> Nicolas Carr argues the human brain can—and is—dynamically rewired as it reacts to information. The way information is presented through Internet technologies, particularly search engines, discourages linear thought and favors a factoid-based way of thinking. Whether our brains are physically rewired is a point for discussion, but the shifts in how we think documented and explained in this book are foundational.

Fernandez, Luke, and Susan J. Matt. *Bored, Lonely, Angry, Stupid: Changing Feelings about Technology, from the Telegraph to Twitter.* Cambridge: Harvard University Press, 2019.

> The authors examine the history of culture and communications technology, starting with the telegraph. Their primary argument is that these technologies don't just change how we feel momentarily, but rather shape the way we feel everything.

Gillespie, Tarleton. *Custodians of the Internet: Platforms, Content Moderation, and the Hidden Decisions That Shape Social Media.* New Haven: Yale University Press, 2018.

Tarleton Gillespie examines and explains many of the techniques used to filter, sort, and promote content posted on social media services.

Jones, Rhys, Jessica Pykett, and Mark Whitehead. *Changing Behaviours: On the Rise of the Psychological State*. Northhaptmon, MA: Elgar, 2014.

The authors trace the rise and implementation of government policies designed to change the attitudes and behaviors of entire cultures and nations. Many of the policies and techniques discussed here are implemented through modern social media systems, both by governments and system operators.

Lovink, Geert. *Sad by Design: On Platform Nihilism*. Kindle ed. London: Pluto, 2019.

Geert Lovink considers the philosophical nihilism of online platforms. Nihilism, in Lovink's context, is the promotion of the self to the position of godhood.

Schüll, Natasha Dow. *Addiction by Design: Machine Gambling in Las Vegas*. Princeton: Princeton University Press, 2014.

Natasha Dow Schull spent time in gambling addiction recovery groups and at gaming conventions, talking to experts in gambling, and even working in the gaming world to discover the tools and techniques gaming companies use to create and maintain the habit of gambling—to the point of addiction. The sections on the user experience and the way gamers are treated as resources are particularly useful.

Postman, Neil, and Andrew Postman. *Amusing Ourselves to Death: Public Discourse in the Age of Show Business*. Harmondsworth: Penguin, 2005.

Neil Postman coined the term *media ecology*, which studies the way media and culture interact with one another. While this book was written before the advent of social media, it contains good discussions on the resonance of media, and how that resonance controls the user experience.

Turner, Fred. *From Counterculture to Cyberculture: Stewart Brand, the Whole Earth Network, and the Rise of Digital Utopianism*. Kindle ed. Chicago: University of Chicago Press, 2006.

Fred Turner outlines the history of the Californian Ideology, starting with its improbable roots in the military-industrial complex and the hippie movement of the 1960s.

Turkle, Sherry. *Alone Together*. New York: Basic, 2017.

> Sherry Turkle investigates the relationships between computers (robots) and humans. The most poignant parts of this book discuss how many people prefer "relationships," such as they are, with computers over relationships in real life. The result of this life is being alone together.

Williams, James. *Stand Out of Our Light*. Cambridge: Cambridge University Press, 2018.

> James Williams explains how social media services consume our attention—and argues these companies need to have more respect for the attention of their users.

Wu, Tim. *The Attention Merchants: The Epic Scramble to Get Inside Our Heads*. New York: Vintage, 2016.

> Tim Wu follows the history of what he calls the attention market from early newspaper advertising through social media networks and modern Internet-based advertising.

Zuboff, Shoshana. *The Age of Surveillance Capitalism*. New York: PublicAffairs, 2019.

> Shoshana Zuboff brilliantly explores and explains the connection between widespread surveillance, the power of social media services, and profit.

Bibliography

"About LinkedIn." https://about.linkedin.com/.

Anticoli, Linda, and Marco Basaldella. "Shut up and Run: The Never-Ending Quest for Social Fitness." In *Companion Proceedings of the The Web Conference*, 1553–56. WWW '18. Association of Computing Machinery, 2018. https://doi.org/10.1145/3184558.3191609.

Arp, Daniel, Erwin Quiring, Christian Wressnegger, and Konrad Rieck. "Privacy Threats through Ultrasonic Side Channels on Mobile Devices." In *2017 IEEE European Symposium on Security and Privacy*, 2017, 35–47. https://doi.org/10.1109/EuroSP.2017.33.

Begbie, Jeremy S., ed. *Beholding the Glory: Incarnation through the Arts*. Grand Rapids: Baker Academic, 2000.

Boers, Elroy, Mohammad H. Afzali, Nicola Newton, and Patricia Conrod. "Association of Screen Time and Depression in Adolescence." *JAMA Pediatrics* 173, no. 9 (September 2019) 853–59. https://doi.org/10.1001/jamapediatrics.2019.1759.

Carr, Nicholas. *The Shallows: What the Internet Is Doing to Our Brains*. Kindle ed. New York: Norton, 2011.

Cummings, R. *Are You Following Jesus or Just Fooling Around?* N.p.: Xulon, 2003. https://books.google.com/books?id=3XzVc1FFkekC.

Donsbach, Wolfgang, Charles T. Salmon, and Yariv Tsfati, eds. *The Spiral of Silence: New Perspectives on Communication and Public Opinion*. London: Routledge, 2013.

Epstein, Robert, and Ronald E. Robertson. "The Search Engine Manipulation Effect (SEME) and Its Possible Impact on the Outcomes of Elections." *Proceedings of the National Academy of Sciences* 112, no. 33 (August 2015) E4512–21. https://doi.org/10.1073/pnas.1419828112.

Farrugia, Rianne. "Facebook and Relationships: A Study of How Social Media Use Is Affecting Long-Term Relationships." MS thesis, Rochester Institute of Technology, 2013. https://www.academia.edu/36070788/Facebook_and_Relationships_A_Study_of_How_Social_Media_Use_is_Affecting_Long-Term_Relationships.

Flanagan, Caitlin. "You Really Need to Quit Twitter." *The Atlantic*, July 5, 2021. https://www.theatlantic.com/ideas/archive/2021/07/twitter-addict-realizes-she-needs-rehab/619343/.

Gattis, Paul. "Facebook Spending Tops $1 Billion at Its Huntsville Data Center Campus." *Al.com*, June 2021. https://www.al.com/news/2021/06/facebook-spending-tops-1-billion-at-its-huntsville-data-center-campus.html.

Google. "Google's Mission, Values & Commitments." www.google.com/commitments/.

BIBLIOGRAPHY

Grasso, Samantha. "The 10 Best Apps for Shooting and Editing Selfies." *The Daily Dot*, June 2016. https://www.dailydot.com/debug/best-selfie-apps-iphone-android/.

Grosser, Benjamin. "What Do Metrics Want?" *Computational Culture*, no. 4 (November 2014) 1–41. http://computationalculture.net/what-do-metrics-want/.

Hammond, Ron, and Hui-Tzu Grace Chou. "Using Facebook: Good for Friendship but Not so Good for Intimate Relationships." In *The Psychology of Social Networking: Personal Experience in Online Communities*, edited by Giuseppe Riva, Brenda K. Wiederhold, Pietro Cipresso, and Aneta Przepiórka, 49–52. Warsaw: De Gruyter Open, 2016.

Hayes, Andrew F., Carroll J. Glynn, and James Shanahan. "Validating the Willingness to Self-Censor Scale: Individual Differences in the Effect of the Climate of Opinion on Opinion Expression." *International Journal of Public Opinion* 17, no. 4 (March 2005). https://www.academia.edu/280701/Validating_the_Willingness_to_Self-Censor_Scale_Individual_Differences_In_the_Effect_of_the_Climate_of_Opinion_on_Opinion_Expression.

Hosie, Rachel. "More people want surgery to look like a filtered version of themselves rather than a celebrity, cosmetic doctor says." *The Independent*, February 2018. http://www.independent.co.uk/life-style/cosmetic-surgery-snapchat-instagram-filters-demand-celebrities-doctor-dr-esho-london-a8197001.html.

Jagannathan, Anand. "Frictionless Sharing: Realizing the Promise of Real-Time Serendipity." *Engage.Social*, May 2017. https://engage.social/blog/social-share/frictionless-sharing-realizing-the-promise-of-real-time-serendipity/.

Korecki, Natahsa, and Eugene Daniels. "'Potentially a Death Sentence': White House Goes off on Vaccine Fearmongers." *Politico*, July 2021. https://www.politico.com/news/2021/07/12/biden-covid-vaccination-campaign-499278.

Laperdrix, Pierre, Nataliia Bielova, Benoit Baudry, and Gildas Avoine. "Browser Fingerprinting: A Survey." *ACM Transactions on the Web* 14, no. 2 (April 2020) 1–33. https://doi.org/10.1145/3386040.

Lewis, C. S. *Miracles*. Kindle ed. New York: HarperCollins, 2009.

———. *Surprised by Joy: The Shape of My Early Life*. Kindle ed. New York: Mariner, 1966.

———. *The Abolition of Man*. New York: HarperCollins, 2009.

———. *The Collected Letters of C. S. Lewis*. 3 vols. Edited by Walter Hooper. San Francisco: HarperCollins, 2004.

Lundak, Marlo. "$1.5 billion Facebook data center planned in Sarpy County one of the largest in the world." *6 News WOWT*, March 2021. https://www.wowt.com/2021/03/25/15-billion-facebook-data-center-planned-in-sarpy-county-one-of-the-largest-in-the-world/.

McDermott, Rose, James Fowler, and Nicholas Christakis. "Breaking Up Is Hard to Do, Unless Everyone Else Is Doing It Too: Social Network Effects on Divorce in a Longitudinal Sample." *Social Forces; a Scientific Medium of Social Study and Interpretation* 92, no. 2 (December 2013) 491–519. https://doi.org/10.1093/sf/sot096.

Miller, Carrie. "How Instagram Is Changing Travel." *National Geographic*, January 2017. https://www.nationalgeographic.com/travel/travel-interests/arts-and-culture/how-instagram-is-changing-travel/.

Osman, Maddy. "Mind-Blowing LinkedIn Statistics and Facts (2019)." Kinsta Managed WordPress Hosting, August 2019. https://kinsta.com/blog/linkedin-statistics/.

Pariser, Eli. *The Filter Bubble: How the New Personalized Web Is Changing What We Read and How We Think.* New York: Penguin, 2011.
Penney, Jonathon. "Chilling Effects: Online Surveillance and Wikipedia Use." *Berkeley Technology Law Journal* 31, no. 1 (January 2016) 117–75. https://doi.org/10.15779/Z38SS13.
Pomerantsev, Peter. "Beyond Propaganda." *Foreign Policy*, June 2015. https://foreignpolicy.com/2015/06/23/beyond-propaganda-legatum-transitions-forum-russia-china-venezuela-syria/.
Postman, Neil. *Amusing Ourselves to Death: Public Discourse in the Age of Show Business.* Introduction by Andrew Postman. Harmondsworth: Penguin, 2005.
Primack, Brian A., Ariel Shensa, Jaime E. Sidani, Erin O. Whaite, Liu yi Lin, Daniel Rosen, Jason B. Colditz, Ana Radovic, and Elizabeth Miller. "Social Media Use and Perceived Social Isolation Among Young Adults in the U.S." *American Journal of Preventive Medicine* 53, no. 1 (July 2017) 1–8. https://doi.org/10.1016/j.amepre.2017.01.010.
Raphael, Rina. "Netflix CEO Reed Hastings: Sleep Is Our Competition." *Fast Company*, November 2017. https://www.fastcompany.com/40491939/netflix-ceo-reed-hastings-sleep-is-our-competition.
Rifkin, Jeremy. *The Age of Access: The New Culture of Hypercapitalism.* Los Angeles: TarcherPerigee, 2001.
Rosen, Larry D., Nancy Cheever, and L. Mark Carrier, eds. *The Wiley Handbook of Psychology, Technology, and Society.* Malden, MA: Wiley-Blackwell, 2015.
Rotondi, Valentina, Luca Stanca, and Miriam Tomasuolo. "Connecting Alone: Smartphone Use, Quality of Social Interactions and Well-Being." *Journal of Economic Psychology* 63 (December 2017) 17–26. https://doi.org/10.1016/j.joep.2017.09.001.
Sagan, Carl. *Cosmos.* New York: Ballantine, 2013.
Schaeffer, Francis A. *Art and the Bible.* Downers Grove, IL: InterVarsity, 2009.
———. *The God Who Is There.* Downers Grove, IL: InterVarsity, 2020.
Schilling, Erin. "Facebook Planning $42 Billion Data Center Expansion East of Atlanta." *Atlanta Business Chronicle*, March 2021. https://www.bizjournals.com/atlanta/news/2021/03/05/facebook-planning-data-center-atlanta.html.
Sifry, Micah L. "Facebook Wants You to Vote on Tuesday. Here's How It Messed with Your Feed in 2012." *Mother Jones*, October 2014. https://www.motherjones.com/politics/2014/10/can-voting-facebook-button-improve-voter-turnout/.
Steers, Mai-Ly N., Robert E. Wickham, and Linda K. Acitelli. "Seeing Everyone Else's Highlight Reels." *Journal of Social and Clinical Psychology* 33, no. 8 (October 2014) 701–31. https://doi.org/10.1521/jscp.2014.33.8.701.
Stone, Linda. "Beyond Simple Multi-Tasking: Continuous Partial Attention." *Linda Stone*, November 2011. https://lindastone.net/2009/11/30/beyond-simple-multi-tasking-continuous-partial-attention/.
Summerville, Amy. "Is Comparison Really the Thief of Joy?" *Psychology Today*, March 21, 2019. https://www.psychologytoday.com/us/blog/multiple-choice/201903/is-comparison-really-the-thief-joy.
Summerville, Amy, and Neal J. Roese. "Dare to Compare: Fact-Based versus Simulation-Based Comparison in Daily Life." *Journal of Experimental Social Psychology* 44, no. 3 (May 2008) 664–71. https://doi.org/10.1016/j.jesp.2007.04.002.

Tams, Stefan, Renaud Legoux, and Pierre-Majorique Léger. "Smartphone Withdrawal Creates Stress: A Moderated Mediation Model of Nomophobia, Social Threat, and Phone Withdrawal." *Computers in Human Behavior* 81 (April 2018) 1–9. https://doi.org/10.1016/j.chb.2017.11.026.

Thaler, Richard H., and Cass R. Sunstein. *Nudge: Improving Decisions About Health, Wealth, and Happiness*. Harmondsworth: Penguin, 2009.

Valenzuela, Sebastián, Daniel Halpern, and James E. Katz. "Social Network Sites, Marriage Well-Being, and Divorce: Survey and State-Level Evidence from the United States." *Computers in Human Behavior* 36 (July 2014) 94–101. https://doi.org/10.1016/j.chb.2014.03.034.

Vardi, Moshe Y. "The Winner-Takes-All Tech Corporation." *Communications of the ACM* 62, no. 11 (November 2018) 7. https://cacm.acm.org/magazines/2019/11/240377-the-winner-takes-all-tech-corporation/fulltext.

Wendel, Stephen. *Designing for Behavior Change: Applying Psychology and Behavioral Economics*. Kindle ed. Sebastopol, CA: O'Reilly Media, 2013.

Wicks, Marshall. "Is It Time to Change? Open Theism and the Divine Timelessness Debate." *The Masters Seminary Journal* 18, no. 1 (Spring 2007) 41–67.

Xu, Fengli, Zhen Tu, Yong Li, Pengyu Zhang, Xiaoming Fu, and Depeng Jin. "Trajectory Recovery from Ash." *Proceedings of the 26th International Conference on World Wide Web*, 2017, 1241–50. https://doi.org/10.1145/3038912.3052620.

Zhao, Shanyang, Sherri Grasmuck, and Jason Martin. "Identity Construction on Facebook." *Computers in Human Behavior* 24, no. 5 (September 2008) 1816–36. https://doi.org/10.1016/j.chb.2008.02.012.

Zuboff, Shoshana. *The Age of Surveillance Capitalism*. New York: PublicAffairs, 2019.

www.ingramcontent.com/pod-product-compliance
Lightning Source LLC
Chambersburg PA
CBHW022119090426
42743CB00008B/926